Sunrise Delayed —
A Personal History of Solar Energy

Frank Kreith

DEDICATION

This book is dedicated to a sunny future.

CONTENTS

PREFACE

In the late afternoon of March 13, 1938 a grimy, dark-green train with seven wagons crossed the frontier from Germany into Holland. There was nothing unusual about the exterior of the train but its cargo was quite extraordinary. It consisted entirely of Jewish children from Vienna between the ages of two and fifteen on their way to England. The average age of the children was less than seven years and each of them carried a small suitcase or backpack filled with all their belongings and had a card board with a number around their necks. At fifteen, I was the oldest of these children. I stood mesmerized at a window trying to digest what I saw: Columns of German tanks and other military vehicles of war were rolling in the direction of Czechoslovakia. I knew that a few weeks earlier the British Prime Minister Neville Chamberlain had met with the German Chancellor Adolf Hitler and agreed to cede the part of Czechoslovakia known as the Sudetenland to Germany. This was an area of Czechoslovakia populated mostly with German-speaking people. But this area also had a chain of mountains that had been prepared as the country's protective barrier against an invasion. With Germany in possession of these mountains, the rest of Czechoslovakia was defenseless. But Hitler had promised Chamberlain that if the Sudetenland were ceded to Germany, this would be his last demand for land because the justification for demanding the lands of Lorrain, Austria, and the Sudetenland had been to join the German-speaking people of Europe to the German Reich. But with the German army marching towards Prague it had become obvious that Hitler's ambitions were to conquer Europe rather than merely joining German-speaking people. Yet, there appeared to be a total denial by the leaders of France and Great Britain that Hitler's ambition was a threat to the civilization of Western Europe.

The train on which I traveled was part of the so-called *Kindertransport*, a program under which children from Germany, Austria, and Czechoslovakia were taken to England, where families had agreed to sponsor the young refugees. The numbers around their necks identified the children.

My sponsors were a wealthy British family by the name of Solomon. The family had a beautiful house and held dinners for some of the wealthy and influential British citizens. Soon after my arrival, I was part of such a dinner and, at some point, stood up and told the guests of what I had seen and how I was afraid that the German war machine was an existential threat to England. My remarks were, of course, uncalled for by the occasion and someone at the table reprimanded me by saying that my fears were only those of a Jewish boy who had been persecuted by the Nazis. My detractor went on to say that England was safe because the country was protected by the ocean and a powerful Navy. I still remember the feeling of helplessness resulting from the total denial of the possibility that a German dictatorship could overwhelm the free and democratic countries of Europe.

When I was beginning to think about writing my autobiography, I remember being stimulated by a similar feeling of helplessness and fear because the world's leadership was in denial of another existential threat—this time in the form of climate change and fossil energy resource depletion. The fossil fuels on which the civilized industrial world was built are running out and the leaders of the developed nations are in denial that this threat is imminent and could have disastrous consequences. Having spent more than half of my life studying our energy future, I feel a similar sense of wanting to shout "CAN'T YOU SEE?" as I did the day I spoke up many years ago at the dinner in England. Although the majority of people in the industrial world are aware that burning fossil fuels is causing global warming and that their availability is finite, unless some drastic action is taken soon, we could face a social upheaval as oceans rise, food supplies shrink, and liquid fuels for our transportation system dry up. Of all the past leaders in the United States, only Jimmy Carter addressed this threat directly when he said, "With the exception of preventing war, this [the energy crisis] is the greatest challenge our country will face during our lifetimes. The energy crisis has not yet overwhelmed us, but it will if we do not act quickly." I am writing this autobiography partly to tell my life story and how I became involved in solar energy, but also to explain what actions we need to take for a peaceful transition to a sustainable and environmentally safe future for our children.

CHAPTER 1—EARLY CHILDHOOD, 1922–1938

The beginning of the 20th century was a golden age for Jewry in Austria. The aging Kaiser Franz Josef the First removed many of the restrictions that previously had made it impossible for Jews to obtain a college education and become professionals in the country. Starting in approximately 1900, the Kaiser began to allow Jews to attend universities, become doctors and lawyers, and participate fully in the economic life of the Austro-Hungarian Empire. However, a few restrictions remained. For instance, in order to be the conductor of the Vienna State Opera, you had to be a Catholic, and Gustav Mahler obliged by changing his religion in order to obtain this prestigious position. For my parents, the Kaiser's actions allowed them the opportunity to become professionals. My mother was one of the first women to graduate from a medical school with an M.D. in dental surgery. My father went to law school and graduated as a full-fledged lawyer who could set up an international practice in the center of Vienna.

My father told me that he had to earn his way through college by giving lessons in Latin and Greek—two subjects that were mandatory in those days. Nevertheless, my father participated fully in the student life of Austria. He joined a Jewish fraternity, which in those days defended its members' honor by dueling with swords. He fought at least one, maybe two duels, and had a long red scar on his left cheek. He also served in the Austrian Army and earned a medal for valor as well as the Iron Cross for wounds suffered on the Russian and Italian fronts in the First World War.

I do not know where or how my parents met. The earliest photo I have of my

parents was taken during the First World War. It shows my father in his Austrian officer's uniform with my mother in a big-rimmed black hat sitting next to him. They appear to have been a strikingly handsome couple.

The story behind this picture is interesting. When my mother received a letter informing her that my father had been wounded on the Italian front she was an intern at the Vienna General Hospital. She was quite concerned that he receive proper treatment in the field hospital and immediately took off to the place from which the letter had been sent. It turned out to be a small village in Tyrol. She was able to find the field hospital and provide extra care for his recovery. The photo was taken in 1917, soon after my father's recovery and return to Vienna.

My mother's family name was Klug, which means clever in German. She was the second oldest daughter in a family of five children—three girls and two boys. Her father had originally been a horse trader in Slovakia. He then moved to Vienna and became a grain merchant on the Austrian stock market. Her grandfather had been a captain on one of the ships on the Danube. The family was not wealthy, but fairly well-to-do. They were very religious and kept a kosher household. All their children, however, deviated from their religious upbringing. The two sons married non-Jewish women and none of the three daughters kept a kosher household. I would say that my mother was basically agnostic, but in her subconscious remained throughout her life very attached to Judaism, its culture, and its practices. This commitment later created some serious problems in the life of her daughter.

I know nothing about my paternal grandfather. He was born in Vienna, but died when my father was quite young. I have been told that he made umbrellas. My father's mother was a seamstress who barely managed to sustain her family, which consisted of my father and his sister, Ellen, who subsequently became a very famous operetta singer. According to what I learned from various relatives, including Ellen, my father was greatly admired by his younger sister. But there is one story that I must have heard more than once: the story of the day his armor as the shining prince was pierced. Ellen had missed school for a day. She was afraid to face the teacher alone and asked her big brother to come with her and give an excuse. My father agreed to help and went with his sister to her school. He knocked on the door of the classroom where a class was in session. At that moment he must have gotten cold feet because he opened the door, gave his sister a shove, closed the door behind her and ran. I do not know what happened to Ellen, but I know she never forgot.

The two families on my father's and my mother's side were friendly, but not close. There were several reasons for this distance. My parents tried very hard to become assimilated in the Viennese society. They resented the influx of Jews from the East, the *Ostjuden*. Most of them were dressed in the traditional garb of their *shtetls* in Poland and Russia: Long black caftans, fur caps called *shtreimel*, and prayer-shawls with the ends (*tzitzits*) protruding. Most of them were poor and depended on

Szeged

government assistance, at least in the early days following their arrival. Many of the Viennese assimilated Jews thought that these immigrants would trigger more anti-Semitism in Vienna.

Grandfather Klug had come from Szeged, a town in the east of Europe that had a large Jewish population before the Second World War. The beautiful synagogue of Szeged shown here is one of the few that survived the Holocaust. He was quite comfortable with the *Ostjuden*, who were often destitute when they arrived. He helped them financially and often gave them breakfast, according to my late cousin Hedy.

My cousin Hedy also told me about a very tragic event that contributed to the lack of closeness of the two families. The oldest son of the Klugs was a womanizer as a young man. One of his mistresses was a young Jewish Polish refugee and when she became pregnant he did not want to marry her. Abortions were illegal in those days and the woman asked my mother for help, hoping that with her medical training she could perform an abortion safely. But my mother was a dental surgeon without any experience in performing an abortion and did not want to risk her future medical career. When she refused to become involved in this dilemma, the woman turned to an illegal back-alley butcher who botched the abortion. The young woman died as a result of blood poisoning. Although this kind of tragedy was unfortunately not uncommon in those days, it shook up our family.

Probably the most important event in our lives was the death of my younger brother Hansi at the age of five. Hansi was two years younger than I and was my father's favorite. I believe that my father never completely recovered from the loss. I recall that both Hansi and I were both very sick. Hansi had diphtheria of the ears, a disease that could today easily be cured by antibiotics. But in those days, the illness required an operation to break the bone behind the ear in order to remove the pus that had formed. In my brother's case, the disease was fatal. I remember having a high fever, hallucinations, and finally, a slow recovery from my bout with the disease. I vividly recall my father taking me in a *Fiaker* (a fancy horse drawn carriage) up the Schőnberg, a small mountain outside of Vienna, in hopes that the fresh air would help with my recovery. I also remember my grandfather Klug saying one day when my father was desolate about Hansi's death, "I wish I could have died in Hansi's place." My

cousin Hedy later told me that my father went to Hansi's grave every evening for many months and read stories or sang songs that used to put Hansi to sleep as a baby. As for myself, I have absolutely no memories of any events prior to my recovery and Hansi's death.

While I was growing up in Austria, I remember that every summer our family would stay with a farmer (*einen Bauern*) who had an extra room, or maybe a small apartment in what is known as the Salzkammergut in Tyrol. This is a beautiful area with blue lakes and high mountains, located just a few hours by train from Vienna. These summers in the Tyrol, or *Sommerfrischen*, as they were called, were memorable. We went hiking in the mountains, swimming and boating in the lakes, and truly enjoyed the outdoors. My father usually joined us at the weekend. His arrivals were great events because he always brought us presents. Sometimes he brought along some of his friends with whom we often went hiking. One particularly memorable trip was climbing a mountain called the "Rax" with another couple named Docent and Mrs. Feiertag. It was a tough hike for me and I was proud to have reached the summit, shown in the picture.

Docent Feiertag was a Jewish high school teacher and his wife was a blond-haired beauty. She was not Jewish, and he survived the war hidden by her in the coal cellar of their house for almost four years. After the war Docent Feiertag was named Minister of Education by the Austrian Chancellor, Preisky. I never saw him and his wife again after our trek up the Rax. In retrospect, I am sure that these wonderful memories of the Austrian Alps contributed to my love for the outdoors and my attraction to Colorado, whose scenery is somewhat reminiscent of that magnificent range of mountains.

The winters in Austria were not severe, and every year during Christmas vacation I had an opportunity to go skiing with students from my high school. In those days there were no ski lifts. The skis were made of wood and had no release bindings. You had to hike up the mountain with the skis on your shoulder, but then could ski down through fresh powder snow. None of us students was a very good skier, but the experience was exhilarating. Two events of those days stand out: One was a New Year's Eve in a small Tyrolean village, where all the inhabitants came out with *Facheln* (torches made with black tar), carrying pots and pans with which they made enormous noise to drive out the evil spirits. We joined them, even though we did not believe in the devil. The other event occurred the day I injured my right knee skiing down one of the Alps. The ski instructor came up on a small sled and took me back to the village holding my right leg under his arm as we hurtled down the mountain. Needless to say, it was both painful and scary. When I arrived in the village, my leg was bandaged up and I was given a crutch to walk back to the train. My parents were very worried when I arrived in Vienna on the crutch, but luckily the injury was not severe and very soon thereafter I was back on skis. Undoubtedly, the memories of skiing down the mountain through fresh powder snow were a great influence in my subsequent decision to settle in Colorado, where skiing was becoming a great attraction.

When I was small we lived in a house on the "Ring" which is equivalent to Central Park West in New York City. The house had a large living room with beautiful paintings and some Biedermeier furniture throughout. But when I was about six or seven years old, the worldwide depression caused my parents' income to shrink and the house on the Ring became too expensive for our family. We moved into a fairly large apartment in an area called Sievering, which is somewhat equivalent to Queens in New York. The apartment was in a four-story apartment house located at Sieveringer Strasse 20, near the yearly wine restaurants called *Heuringen,* and was large enough so that my mother could have her dental office in one of the rooms. After we moved, I was enrolled in a grammar school near our house in Manergetta Gasse. It was close enough so I could walk to this school, which I attended for four years. In the winter I could use my wooden skis. I still have a photograph of our class and the imposing woman teacher.

Behind the house were barracks left over from World War I. In these barracks lived poor, mostly unemployed or under-employed people who barely scraped out a living. My best friend was the son of a taxi driver who lived in one of these barracks. His name was Ossie, and we played Indians and Trappers together. Our hangout was a relatively undeveloped piece of ground across the street from our apartment on which my father had started building a small wooden hut. He must have run out of money, because the hut was never finished. It served, however, as a wonderful place to hide or to pretend it was a fort where we would play settlers who had to fend off the attack of Red Indians. Our model for these games were the books of an author by the name of Karl May. He wrote several books about the American frontier, always with two characters. One was a trapper by the name of Shatterhand and the other an Indian named Winnetou. Karl May had never been to America, but his imagination was even better than the real frontier life. His books at the time reflected a phenomenon somewhat similar to the contemporary Harry Potter books by J.K. Rowling.

Most of the people living in the barracks behind our house at that time were Communists. I remember Ossie telling me that his father participated in a strike for better wages for taxi drivers and subsequently lost his job. The situation in his family became dire, and he told me that his 17-year-old sister went out in the evening and met with men who gave her family some money out of friendship. Neither Ossie nor I understood at the time what her nightly activities really meant. He said his sister dressed fancy in the evening, but for our Indians and Trapper games, she came in her usual street clothes.

Soon after we moved in, my mother set up her dental office in our apartment. Our entry hall doubled as a waiting room for her patients. She had no assistant and her dental equipment was fairly primitive, consisting mainly of hand tools and a dental drill operated by a foot pedal. She helped the unemployed or under-employed people living behind our house by doing their dental work free of charge if they could not pay. I do not believe that her practice was very extensive or that she earned a great deal of money, but she was highly respected and much appreciated by our less-fortunate neighbors.

My father had his *Kanzelei* (office) in downtown Vienna. We did not own a car, but a streetcar ran in front of our apartment house and my father always used it to get to work. He shared the office with at least one other attorney. My father's specialty was international law, but he also handled other cases. His largest clients were Siemens & Halske, a German radio manufacturer, and a large Italian insurance company, Generali Italiano. I recall that he usually came home for lunch and then took a short nap on a couch in the living room. The streetcar he used, number 39, was still running when my wife and I visited Vienna almost 70 years later, and it had the same colorful carriages that I remember from my youth.

I did not have a formal religious upbringing. But my father had a lot of Jewish friends, and he was therefore anxious to show them our family's pride in its Jewish heritage by having me become a Bar Mitzvah. I obliged by taking lessons in reading Hebrew and reciting sections of the Torah for the occasion. But I read without comprehending what the words meant and did not develop any real appreciation of the richness of our Jewish heritage and religion. The Bar Mitzvah ceremony was more of a social occasion than a religious one, but I did what was expected of me and my father was pleased. The other part of my religious background was the compulsory two hours per week of religion in all Austrian schools. The instructor was a middle-aged Austrian rabbi by the name of Murmelstein. He, too, merely asked his students to read Hebrew without trying to teach the meaning of what we read or what our religion had to offer in a historical or philosophical sense. I think it was simply a job to him, not an opportunity to teach young boys about their ancestry and religious history. After the war, it was rumored that the man had actually cooperated with the Gestapo in organizing the transport of Viennese Jews to the extermination camps in the East. I also was told that at some point he lost his usefulness to the Nazis and was himself put into one of the cattle cars that the Germans used to ship the victims to the death camps. Murmelstein, however, never completed the trip. He was beaten to death by some of his fellow victims who knew of his previous collaboration with the Gestapo.

The political situation in Austria became precarious after the Depression of the 1930s. The Socialist government elected after the defeat of Austria in the First World War in 1918 was replaced by a mildly Fascist one that clamped down hard on Communists and Socialist blue-collar workers. However, there remained a basic social network that kept people from being hungry and, in most instances, also from being cold. There were soup kitchens all over Vienna that dispensed food, but there were rumors that the food was barely edible. The semi-tranquility of the period came to an end in February of 1934, when the Communist and Socialist Parties went on a national strike. In reaction, the Prime Minister formed an alliance with Graf Starhemberg, the leader of the Austrian Fascist militia, who brought up artillery and bombarded the apartment houses in our neighborhood in which the "Red" families lived. I remember my father asking us to sleep right under the window in order to protect us from stray bullets fired by the Army besieging the buildings that had been built in the 1920s by the Socialist government for Viennese workers.

The next big event in Austria was an attempted *Putsch*—a sudden violent attempt to seize power—by the Nazis in July of 1934. A small group of Nazi shock troops gained control of the Viennese radio station and captured the diminutive Prime Minster, Paul Dollfuss. The Prime Minister was injured in the process and the Nazis held him as a hostage in the main government building, where they had retreated after their *Putsch* failed. Dollfuss was bleeding from a gun wound, but the Nazis did not permit a doctor to treat him. As a result, Dollfuss bled to death and, after a day's standoff, the Nazis finally surrendered. They were captured and the Austrian Supreme Court sentenced them to life imprisonment. However, they were freed and treated as heroes after Hitler annexed Austria four years later.

Economic conditions in Europe had gone from bad to worse after the stock market crash of 1929. Hitler had gained power in 1933 in Germany and soon began a campaign to join all German-speaking people in a single nation. He demanded that Austria become a part of the Greater Germany. In response to Hitler's demand for an *Anschluss*, the Austrian Prime Minister decided to hold a plebiscite in which the Austrian people could decide whether they wanted to remain independent or become part of Germany. The election was set for Sunday, March 12, 1938. Two or three days before, many thousands of people were out in the streets of Vienna protesting against the *Anschluss*. I, as well as many of my friends who were naturally opposed to the *Anschluss*, joined the protesters. Although in some parts of the city there were terrible fights, the majority of people was peaceful and seemed opposed to the idea of becoming part of Hitler's Germany. On Saturday, the day before the referendum, we were still hopeful that people would vote against the *Anschluss*. But Hitler took matters into his own hands and, in the early hours of Sunday, the day set for the election, the German troops marched into Austria with virtually no resistance from the government. The Prime Minister at the time, Kurt von Schuschnigg, fled to England, while many Austrians cheered the arrival of Hitler's army in the streets and waved flags with red swastikas. Since the other European powers quietly accepted Hitler's takeover of Austria, this event was probably the first step toward a major change in the power structure of Europe.

CHAPTER 2—AUSTRIA AFTER THE NAZI TAKEOVER, 1938–1939

On March 12, 1938, I remember looking out of the window from our apartment in the late afternoon when I heard goosesteps below. There I saw a contingent of SA troops marching past in formation while singing their famous fighting song with the words: "*Heute gehoert uns Deutschland und morgen die ganze Welt*" (today we own Germany and tomorrow the entire world). We turned on the radio and heard that the German Army had crossed into Austria and that Prime Minister Schuschnigg had abdicated in favor of a Nazi lawyer by the name of Seyss-Inquart. The news was quite a shock, but was not totally unexpected in view of Hitler's demand for an *Anschluss* and the unwillingness of European countries to take a stand against the German dictator. We had heard bad news from Germany and knew that many Jews had emigrated and that many more were trying to leave. On the other hand, we did not know what the *Anschluss* would mean for Austria and Austrian Jews in particular.

Soon after we heard on the radio that the German troops had crossed into Austria, my father, whom we called Ope, told us that he would visit his sister and left the house. Ope's sister was named Ella, but she always insisted on being called Ellen and never Ellie or Ella. She was married to a man considerably older than she named Heinz Reichert, who was famous in the musical circles of Vienna. He was well known because he had written the librettos for many of the Viennese operettas by Franz Lehar and Erich Korngold, as well as for some of the famous Italian opera composers, including Giacomo Puccini. Heinz's original name was Blumenthal, but sometime during his life, he decided to change it to the less Jewish-sounding name Reichert. Heinz was married to a Catholic woman when he met Ellen. She was, at that time, a well-known singer who appeared in many of his operettas. I do not know much about their early life; but from later events, I gathered that Ellen had been Heinz's mistress for many years because his first wife would not give him a divorce. Ope helped Ellen in this difficult situation and arranged to have her become the legal heiress to Heinz's fortune. Eventually he was even able to make it possible for Heinz and Ellen to marry.

Heinz was a well-to-do man in Vienna and the two lived a sumptuous life with an automobile and a home in one of the finest parts of the city. My parents and the Reicherts did not get along well for at least two reasons: First of all, my parents' lifestyle was at the level of lower-middle class, whereas the Reicherts were part of the upper echelon of Viennese society. But more importantly, for reasons unknown to me, Ellen hated my mother with a vengeance. I never learned the reason why, but suspect that Ellen was envious of my mother's high status as a "Frau Doktor."

The evening of March 13th, Ope came home and told us that had he urged Heinz and Ellen to leave the country as quickly as possible because Heinz's reputation and previous marriage made him vulnerable to accusations of racial misconduct from the Nazis.

Heinz and Ellen had a beautiful touring sedan, and apparently took Ope's advice seriously. They left Vienna that same night and crossed into Switzerland the next day before the frontiers were closed by the Germans. Switzerland had always been a haven for people who wanted to put money in a safe place, and apparently Heinz had stashed away some of his fortune in a Swiss bank. Moreover, they had a lot of friends in the musical business all over the world. Some of them had previously gone to Hollywood and were able to send Heinz and Ellen an affidavit that allowed them to immigrate to the United States. Soon thereafter, the *Geheime Staatspolizei* (Gestapo) of the Nazi government of Austria issued a decree confiscating all of Heinz's belongings in Vienna.

60088

Geheime Staatspolizei
Staatspolizeileitstelle Wien.
Referat II B J 3 A

Wien, den .... 194...

19.7.

1

B.-Nr. 1289

Beſchlagnahmeverfügung.

Das geſamte ſtehende und liegende Vermögen ſowie alle Rechte und An-
ſprüche des (der)

Reichert Heinrich Iſr. (fr.Blumenreich)

geb. am .... **27.12.1877** .... in .... **Wien**

und ſeiner Ehefrau .... **geſch.,**

geb. am .... in ....

ſowie

deren Kinder ....

zuletzt wohnhaft geweſen in Wien .... **, 13. Auho ſtr.68**

Staatsangehörigkeit .... **DR.,**

wird aus Gründen der öffentlichen Sicherheit und Ordnung mit dem Ziele der ſpäteren
Einziehung zu Gunſten des Deutſchen Reiches beſchlagnahmt.

Dieſe Beſchlagnahmeverfügung erliſcht ohne formelle Aufhebung mit dem Über-
gang der Eigentumsrechte auf das Deutſche Reich. Die Verfallserklärung zu Gunſten
des Deutſchen Reiches wird im Reichsanzeiger verlautbart.

Ein Rechtsmittel gegen dieſe Beſchlagnahmeverfügung iſt nicht zuläſſig.

Geheime Staatspolizei
Staatspolizeileitſtelle Wien.

IA   gez.   **Dr. Ebner**

Many years after the war, I learned that Uncle Heinz had left instructions with
one of his musical coworkers, a Jew by the name of Leo Stein, to keep his memorabilia
safe. Leo and Heinz had worked together on librettos for a number of operettas, but
Leo was best known for having written the words to the well-known Viennese operetta
called "The Merry Widow," which was based on music by the famous Austrian
composer Franz Lehar. My younger brother Kurt frequently traveled to Vienna in later
years. He told me that Leo, although he was Jewish, had been protected from Nazi
persecution by Hitler personally because Hitler loved to listen to "The Merry Widow"
and often surreptitiously sneaked into Vienna when the opera was performed at the
Vienna State Opera (*Wiener Staats-Oper*). Leo continued to live under Hitler's

protection even after the *Anschluss* and the start of the war. He passed away in 1943 and thus, unfortunately, did not live to see the end of the war or the downfall of Hitler. Leo, however, obviously took Heinz's instructions to heart and took good care of the memorabilia that Heinz left behind. After the war, the material commemorating Heinz Reichert's work was incorporated in one of the national museums, where it remains to this day. I have never been to that museum, but my brother tells me that he visited it on one of his trips to Vienna.

I have often wondered what would have happened to my parents' lives if Ope had himself acted on his own advice. Ope had always been an ardent Zionist, yet he never had the courage to act on his beliefs and leave Europe for Palestine, as it was known at that time. Of course, Ope had three children and did not, as far as I know, have any money outside of Austria. Emigration to Israel would have been much more difficult for him and his family than for his childless, wealthy sister. All the same, I think a move to Israel might have given him a chance for a more meaningful life that would have maintained his pride and self-esteem. But after Ope returned home that night, he was still hoping that the Nazis would look at him as a loyal Austrian who had fought for his country during the First World War, rather than simply as a Jew who would have to be driven from his position as lawyer. Of course, he was not the only Jew who soon realized that being a native Viennese would count for nothing in the eyes of the Nazi bureaucracy.

Hitler made his triumphant entry into Vienna on March 13, 1939. A photo commemorating his entry shows an Austrian civilian by the name of Arthur Seyss-Inquart standing next to Hitler. Seyss-Inquart was a lawyer, the same age as my father, with whom he studied law at the University of Vienna. He was a known anti-Semite, but for political reasons did not join the illegal Nazi party. This made it possible for him to take over the chancellorship after Kurt von Schuschnigg resigned on March 11th. The next day, German troops crossed into Austria and Seyss-Inquart conveniently sent a telegram to Hitler inviting them to come. Two days later, he drafted legislation making Austria a part of Germany called Ostmark. Seyss-Inquart later became the administrative chief of southern Poland, where he arranged for the Jews' movement into the ghetto and subsequently to Auschwitz. After Hitler committed suicide in his bunker, Seyss-Inquart attempted to form an interim German government. But his attempts were thwarted when he was arrested on May 7, 1945, by two British soldiers in Hamburg. It seems like poetic justice that one of the soldiers was Norbert Mueller, a young German Jew who had escaped from Austria on a Kindertransport, just as I did. Seyss-Inquart was convicted of crimes against humanity at the Nuremberg trials and was hanged on October 16, 1946.

The first signs of the takeover became apparent the next day, when many of the people that we had known for a long time proudly displayed a swastika pin on their coats as an indication that they had been members of the illegal Nazi party. Many of the people that suddenly showed their Nazi sympathies had previously been quite friendly with us. But after the takeover, a certain distance between Christian and Jewish Austrians became apparent. When we walked out into the street or into a grocery store, a lot of people greeted each other by raising their right arm and saying "Heil Hitler," instead of the conventional "Grüss Gott" that had been used as a greeting in Austria for centuries. For Jews, this Hitler greeting was of course forbidden.

There were no immediate repercussions on our lives, but as the days went on my father told us about one client after another informing him that they were transferring their business to a non-Jewish lawyer. Some of the patients that my mother had taken care of informed her that they would no longer use her services. I was shocked to see how many of the former Communists living in the barracks suddenly began to wear swastikas, although these were different from those that former illegal members of the party had on their lapels. A real shock came to me when my bosom-friend Ossie told me that he was joining the Hitlerjugend (Hitler Youth). I asked Ossie how he felt about the anti-Semitic position and actions of that group. His response was, "Well, Franzl, I like you, but I am told that the Jews are an inferior race of dirty money lenders and that one cannot make an exception about one's attitude because one or two Jews you know are not a part of that nasty group." I asked him if he knew many people who belonged to that nasty money-grabbing group. He said, "I do not know anyone personally, but the people living in the barracks tell me that the reason that they are poor and unemployed is because of the Jews." Sadly, our days of playing Trapper and Indian soon came to an end.

B
A
L
L

DES
DÖBLINGER GYMNASIUMS

I was, at that time, attending the Döblinger Gymnasium as a student in the sixth grade. For the first few weeks we continued to attend classes, take tests, and interact with others as usual. The figure shows the drawing of our school advertising a dance. The first break came when our gymnastics teacher, a man named Kolmarek, assembled all the students in the *Turn Halle* (gymnasium) and asked that "those who are Jewish should step forward and then march over to the

13

right." After he had separated the Jewish students, he informed us that henceforth we would not be doing any exercises with the Aryan group. I realized at that time that this would be only the beginning of other measures taken against Jewish students. Indeed, I was right, because very soon the entire school system would undergo a radical change.

After the Jewish students were separated from their Aryan classmates for gymnastics classes, some of us began to form a small group that would exercise on our own. But the separation for gymnastics classes was soon followed by a much more severe and far-reaching edict. On April 29, 1938, the officials of the Education System in Vienna sent down an edict that all of the non-Aryan students (we were no longer referred to as "Jews," but simply as "non-Aryan") would be expelled from the Gymnasium and sent to another school set aside for non-Aryan students. Starting the following Monday, 104 students from the Döblinger Gymnasium, including me, went to the new school to which we had been assigned.[*] It was a dilapidated place and there were obviously no serious plans for continuing our formal education. Some Jewish teachers had been assigned to the school, but it was clear that from now on we would be a group of second-class citizens. The most amazing aspect of this separation was that a number of students who had regularly attended Catholic or Protestant religion classes, which were mandatory for Christian students, also appeared at the non-Aryan school. We were told that these students were not of Jewish religion, but of the "Hebrew race" according to the newly instituted Nurnberg racial laws, which required a person to have more than three-quarters of non-Jewish blood to be considered Aryan. Many of these students had heretofore been unaware that they were not "real" Catholics or Protestants, and they were the most humiliated of all of us. The most stunning part of this development was that none of the Catholic priests or Protestant reverends in Austria raised their voice in protest. In fact, the Austrian Cardinal Initzer openly proclaimed his support of Hitler, as well as of his racial policies. I never had much faith in the Christian edict to "love thy neighbor as thyself," and this proclamation of the Austrian leader of the Catholic Church only confirmed my feelings.

---

[*] A book documenting the fate of all 104 students was published in 1999 under the name *Vertreibungsschiksale* by Martin Krist, and published by Verlag Turig & Kant.

A photograph of the students in my sixth year Gymnasium class prior to segregation. I am second from the right in the second row. Otto Holzapfel is the first on the left and Ossie is first on the right in the first row.

In the background is Professor Hruby, a violent opponent of Hitler, who was dismissed soon after the Anschluss in 1938.

After having been expelled from the gymnastics exercises at the Dőblinger Gymnasium, a group of six students began to form a clique or underground cell. We began to meet regularly and take stock of our situation. It was clear to us that to survive we had to leave Austria as soon as possible, and each of us began to form his own plan of escape. The six of us were Wolfi Pollack, who was a talented artist; Robert Silberstein, who wanted to be an opera singer; Paul Toch, who wanted to become a doctor; Rudi Bunzl, who wanted to become an engineer in the factory of his parents; a big and brawny young man by the name of Kurt Huppert who had been brought up as a Catholic, but whose parents had Jewish ancestry; and myself. I was anxious to leave but did not really know how and towards what purpose. None of us was particularly religious and the bond that held us together was our common fate and the exclusion from other students our age.

CHAPTER 3—*KRISTALLNACHT* (THE NIGHT OF BROKEN GLASS), NOVEMBER 9, 1938

"Human beings can be awful cruel to one another"
-Mark Twain, The Adventures of Huckleberry Finn

The semi-normal life that my family and our friends led despite the restrictions imposed by the Nazi regime came to a sudden halt on November 9, 1938. The day before, a young Jew had shot a Nazi official in Switzerland who died later that day. This gave Hitler the long-awaited chance to clamp down on Jews and take away their property. We heard about the assassination on the radio and just two or three hours later a group of local Nazis, including many of the former Communists from the barracks, came to our apartment and asked that I come out and help them "clean benches" in one of their meeting halls across the street. I was obviously pretty scared, but followed the leader of the motley gang who then asked me also to scrub the street in front of their meeting hall. At that point, my mother rushed out of the apartment

and confronted the Nazi group, saying that she would take my place for the scrubbing of the street. It was a pitiful moment, because many of the Nazis confronting us had formerly been her patients, and quite a few of them had been treated for free by my mother.

When my mother rushed into the street and offered to take my place in scrubbing the benches and the sidewalk, I imagine that she wanted to help her son, and possibly also shame the people for whom she had provided free medical help into releasing me. She obviously expected them to refuse her offer, but the reaction of the mob was quite different. They yelled, "The Jew doctor wants to help. Let her clean our streets for a change." As for me, I was terribly embarrassed by my mother's actions and would have much preferred to continue cleaning than to see my mother be part of this shameful demonstration of Nazi power. I think we got off quite lightly, because I merely suffered a few kicks from Nazi boots on my behind and my mother was just shoved into a bucket of dirty water by one of the onlookers. After about two hours, the Nazis got tired of watching old women cleaning the sidewalk and released us. We were able to go home, but life was not the same afterwards. A sense of fear and foreboding became a part of our life.

The next day when we walked down Sieveringer Strasse, we saw the destruction which the Nazis had wreaked the previous night. When we came to the synagogue, we saw that the Nazis had broken all of the windows, smashed the furniture, and torn into pieces the Torah, which is the sacred book of the Jews. We also saw that the Nazis had destroyed all of the stores owned by Jews along the street and written, with red paint, the word *Jude* (Jew) across the store windows or on the entrance doors. These events had been repeated all over both Germany and Austria. In retrospect, this night has been appropriately called *Kristallnacht*—the night of broken glass. The events of this night only reinforced our knowledge that worse things were still to come, and our only recourse would be escape.

A day or two later, another sad incident occurred. Some of the inhabitants of the barracks, some our neighbors, accompanied by an SA officer, knocked on the door of our apartment and asked to be admitted. My father must have been preparing for such a visit and brought out the medals that he had earned fighting for Austria, the saber that he had used as an officer in the Austrian Army, and the helmet he had worn during battle. He thought that these items would somehow sway our visitors and deter them from their goal, which obviously was to demean a Jewish family. However, they paid no attention whatsoever to my father's demonstration of past Austrian loyalty and simply asked that he hand over some of the essentials of his profession, including a typewriter, and some of our household goods, including the silverware and some of the art that he had collected over the years. The art included some paintings, porcelain figures, and engraved copper plates that he had collected. One of the neighbors, acting as an official, took out a form prepared by the Nazi Government and wrote down all the items that were confiscated from our household. I felt great pity for the way my father was treated by the mob and how his self esteem must have suffered from the futility of demonstrating his loyalty to the country. On the positive side, however, from that day onward he no longer opposed my efforts to leave Austria at any price.

The day after Kristallnacht, many Jews were arrested and taken to one of the concentration camps called Dachau. My uncle, Walter, the husband of my maternal Aunt Ida, was included among those taken, as was my friend Paul Toch's father. Although some of the prisoners were tortured and murdered in captivity, Dachau was not an extermination camp like Auschwitz, but rather a place to hold and scare Jews and political prisoners into accepting the Nazi rule. The condition for release of the Jews was usually that the family obtains some kind of exit visa, as well as proof that they could pay for the transportation necessary to immigrate. In my uncle's case, his wife Ida was able to obtain a visa to Australia and he was released from the concentration camp. He had been beaten and tortured, but his overall health was not broken. Similarly, Paul's father was released after two weeks at Dachau. He too had been beaten, deprived of food and drink, but also survived the ordeal. Paul's family was fortunate enough to get a visa to immigrate to the United States where they had some relatives. But many other Jewish families were not that lucky. Despite the knowledge of the plight of the Jews in Germany and Austria, very few countries would provide visas for Jews to escape and emigrate unless they had money or relatives in that country. For many, the situation became so hopeless that a large number of them reverted to the only escape available to them—the final peace offered by allowing the gas from their kitchen stoves to fill their homes.

CHAPTER 4—ESCAPE FROM VIENNA ON A KINDERTRANSPORT, MARCH 13, 1938

The small cell of six that I have spoken of before continued to make plans to escape. The first one to try was Robert Silverman. He was an excellent skier and attempted to cross the border from Austria to Switzerland. I do not know any details, but we learned that the border patrol caught him in the process and when he failed to stop, shot him in the back and killed him. Wolfi Pollack's father was an engineer who was able to get a visa to go to Albania in order to help the somewhat primitive Muslim country build a railroad. Wolfi sent us postcards on which he sketched the life around him. In particular, I remember the Oriental shoes of the Albanian men with their up-curved tips. Life must have been very hard for Wolfi because he did not speak the language, and the Muslim natives were not friendly to Jews. I learned subsequently that Wolfi went to France where he joined the French Underground. But after he left Albania, the postcards stopped and I heard no more about his fate until recently, when documents of the people killed in the extermination camp at Auschwitz were released. Among these documents was an entry that Wolfi had been captured in France while fighting for the Underground. He was transported to Auschwitz where he was gassed. The Bunzels, who had been one of the wealthiest families in Europe, were able to emmigrate to London where, after my own escape, I subsequently met up with my friend Rudi. The last of the six of our cabal, Kurt Huppert was not Jewish. He had been an active member of the local church before the Nazi occupation. However, his birth certificate showed that he had two Jewish grandparents, which made him a Jew under the Nazi's Nuremberg race laws. Kurt was also able to escape to England. Another student shown in the photo of my class, Otto Holzapfel, emmigrated directly to the United States where he became a famous Professor of the History of Science at Harvard under the name of Gerald Holten. He has recently published a book on what he called the Second Wave of Immigrants. It is an account of the life of children who were refugees from Hitler. I contributed some of my life experiences to that book.

Two of my uncles had gone to Australia with their families; the third had emmigrated to Shanghai, which was one of the few places that permitted Jews to enter. This uncle's wife, who was Catholic, left Vienna after her husband had arrived in Shanghai, ostensibly on a pilgrimage to the Vatican. As an Aryan Catholic woman, she was permitted to go to Rome. But, on her journey she smuggled a valuable diamond out of Austria, managed to sell it in Italy, and subsequently joined her husband in Shanghai where they opened a bar in the International Zone. I really had no specific prospects for escape, but I remember saying to myself, "If I ever get out of this, I do not want just to exist, but I want my life to count for something." I have carried this commitment with me for the rest of my life.

After the *Anschluss*, the center for information about what was happening to the Jews of Vienna was a place called the *Kultusgemeinde*. It was a place where information was exchanged, both formally and informally, regarding prospects of escape, advice about schools, opportunities for learning new skills that could be of use in making a living in another country, and hearing about the fate of people who had been taken away to concentration camps. I went to that center twice every week without actually having any specific goal in mind. But one day, when I looked at the board on which information about various topics was posted, I saw a notice regarding a new program entitled *Kindertransport*. Upon reading the description of the program, I saw that here was my opportunity to escape.

Although the British government forbade the immigration of Jews, claiming that unemployment was too high, in a rare act of mercy that began in December of 1938 and ended with the outbreak of war nine months later, Britain allowed Jewish children from Germany, Austria, and later also Czechoslovakia to immigrate, provided someone guaranteed that the child would not become a burden to the British government. The U.S. government refused to pass a similar bill to accept 20,000 unaccompanied children on the grounds that it was cruel to separate families. It is hard for me to accept this rationale, because as cruel as it obviously is to ask parents to send away their children, it was the only way for many to save the lives of their offspring. When I saw this opportunity to emmigrate to England, I rushed home and asked my father to contact his sister to find me a guarantor in order to leave Austria. At the same time I applied to the agency responsible for selecting the children to leave Austria on the *Kindertransport*. A few weeks later I learned that my aunt had been able to find a guarantor in England. I believe she contacted a former agent of her husband's by the name of Joe Fenston, and he subsequently found the Solomon family who became my official sponsors. Although it must have been a hard decision for my parents to let their 15-year-old son go unaccompanied to a new country, I had no hesitation in applying for the first opportunity to leave Austria. This opportunity came on the 12th of March, 1939—almost exactly one year to the day following Hitler's annexation of Austria. I had no doubt that it was a great sacrifice for my parents to put me on a train with the uncertainty of whether or not they would ever see me again. But they must have become aware of the need to make this sacrifice, because a few months later my parents also sent my little sister Susi, who was nine years old at that time, on a transport to England.

When I arrived at the Vienna railroad station on March 12, 1939, I was just a little over 15 years old. A railroad station in those days was a dismal place. The locomotives belched black smoke from burning coal, and that smoke adhered to the inside of the station's glass enclosures, making the railroad station appear grey, even when the sun was shining outside. It was early morning when I arrived at the station, but it was already teeming with two or three hundred children ranging in age from three to sixteen years. This was a day when one of the first of many Children's Transports was leaving Vienna for London—a desperate effort designed to save children of Jewish ancestry from the persecution of the Nazi regime.

I did not know any of the children. Most of them were accompanied by one or more adults, many by their parents. The one thing that all of us had in common was a piece of cardboard around our necks containing a number. That number was to identify each child to the unknown foster parents in London who had offered to accept one of the refugee children under the auspices of the *Kindertransport*.

All of us were urged to enter the train as quickly as possible. Although a sense of fear of the unknown future was in the air, there was no panic among the children. On the train were two adults who tried to keep order among the children, some of whom were crying for their parents, while others were trying to protect their belongings by stashing them in a corner or in a net above the seats. Under the regulations of the Children's Transport, each child was only allowed to take one piece of luggage. I had a knapsack, and in it I had some clothes, an extra pair of shoes, a raincoat, and a tin box that I had used in Vienna to carry my lunch to school. On my arm I had a steel Omega watch, which my father had bought for me a few days before departing. My father owned a beautiful gold pocket watch, but under the conditions of the Transport we were not allowed to take along any money, jewelry, or gold. Hence, my father thought it would be wise to buy me a good steel watch that would serve its purpose, but would not create a problem for me in case a German Storm Trooper searched my body and luggage. It was a wise decision, because when we reached the border the train stopped and Nazi SA troopers with large swastikas on the sleeves of their uniforms came on the train and searched each of the children and their luggage for contraband goods. They found none in our compartment, but our hearts were beating loudly while the search was on. After an hour so, the train began to move again.

The rest of the journey was really uneventful, except for the crying of the younger children who were missing their parents or caretakers. In my compartment, there was a three-year-old who continuously screamed about her experience of a few nights ago when Storm Troopers had entered their flat and one of them had put a knife into the heart of her father. I was probably one of the oldest in this group of children, and was asked by some what to expect. But all I knew was that London was a big city and that the British were kind to refugees from the Nazis. The bobby welcoming the children was my image of freedom.

It was a dreary journey because all of us were facing such an uncertain future. There was hardly any food or drink, and all of us were anxious about what would happen at the border. We all breathed a sigh of relief when the train began to move again, crossing into Holland after the long and tedious trip across Germany. On the other side of the border, the train stopped again and there were women with hot chocolate and cookies waiting for us. It is hard to describe the feeling of being again among kind human beings. I called the cup of chocolate offered me *Meinen Freiheits Trunk* (my freedom drink). After this brief stop, the train continued to the coast of the North Sea, arriving in the evening at the so-called Hock, which is the terminal for the ferry between Holland and England. I had never before been on a large ocean vessel, nor had I experienced the queasy feeling in my stomach that began when the waves caused the ferry to sway. But, the elation and expectation of a new life soon overcame the discomfort of the journey. The boat left at night, and after a few hours we landed, boarded another train, and finally arrived in London. It was the morning of March 14th and the headlines in the *London Times* said that Hitler had crossed the German frontier the night before and occupied the remnants of Czechoslovakia. A few months before, Hitler had met the British Prime Minister Chamberlain in Berchtesgaden where he promised that if the Czech province of Sudetenland, which had a large German-speaking population, were ceded to Germany, he would make no more demands for other lands. Chamberlain returned to England and announced that this agreement would assure the world of "peace in our time." I realized that this invasion meant, of course, that the promise Hitler had given the British Prime Minister Chamberlain for peace had been shattered and that war had become inevitable.

Many years later, the British government realized the significance of the *Kindertransport* in saving the lives of thousands of Jewish children from Nazi extermination and erected a statue to remind the world. The statue stands at the Liverpool Station, where the children of the *Kindertransports* arrived in England.

In early 2000, I received a letter informing me that a movie had been made about the *Kindertransport*. The letter came from a group of former kindern in England who arranged yearly meetings to remember the experience of escape from Nazi Germany. A few days later I had the opportunity of viewing the film, *Into the Arms of Strangers*, and also learned that it had won the prestigious Academy Award for Best Documentary Feature for the year 2000. The film was largely based on interviews with several of the children who had been saved through the Children's Transport

and had been taken in by foster homes in England. It was beautifully narrated by the award-winning actress, Dame Judi Dench. I was somewhat disappointed that neither my sister Susi nor I had been interviewed for the film, but happy to see that the story of the 10,000 children who had escaped with the *Kindertransport* had now been committed to memory by this film. The film was unfortunately not a box-office success; possibly because it had only a limited distribution despite the recognition it received by the Academy.

CHAPTER 5—LIFE IN LONDON, MARCH – JULY, 1938

Once we arrived at the London station, the waiting foster parents scrambled to find the number on a child that could identify which of the children belonged to whom. Being taller than the rest, I had no problem being identified by the family that had offered to accept me into their home. Their name was Solomon. They were wealthy Jewish people by my standards. They owned a big house, a car, and had two servants. The family consisted of Mr. and Mrs. Solomon, both in their late 40s or early 50s, two daughters of about ten or eleven, and a son in his early 20s. The son had a badly disfigured face as a result of a horse having kicked him some years previously. It was a luxury to get into their car and be driven through London into a garden area called Kensington where the Solomon's house was located. I only spoke a very few words of English and communication between us was minimal. I tried very hard to anticipate what was expected of me, but with limited verbal communication, this was difficult and I was tense and uneasy.

When we arrived at the Solomon's residence, I was introduced to the maid and given very specific instructions about what time breakfast was served, when one had to be ready for dinner, and what clothes one was expected to wear. The Solomons also told me that they would expect me to enroll in the Polytechnic Craft School—a sheet metal trade school—which would teach me enough about metal production that I would soon be able to get a job and earn my own living. I attended the school on and off from April until June. I do not remember very much about the sheet metal school, because I was really not interested in learning about how to bend aluminum sheets. But, I did learn how to read blueprints, something that would be exceedingly useful to me in the future. I also picked up a good deal of English vocabulary dealing with machines, metal production, and British snobbery. Since I did not like the school very much, I often played hooky in the afternoon and did only the minimal necessary to survive.

As soon as I arrived in London I started to look for an opportunity for my sister and my brother to follow me on one of the children transports. I was only successful in finding a family to provide guarantee for my sister Susi to leave. Susi was a lovely nine-year-old girl with a lot of freckles and blondish-red hair. I had taken a photo of her with me and showed it to all the prospective people I came in contact with. The Jewish agency was the one who learned about a physician's family interested in taking in a little girl and I showed the family Susi's picture. After some hesitation, because they really wanted a baby, the family agreed to sponsor Susi. She came on a transport in late summer and went to Brighton where the physician and his family lived. She immediately started going to school and was already very anglicized when I was finally able to visit her at Christmas.

It was spring in London when I arrived and I loved the city. There were beautiful gardens, places where people would play, and a fascinating place called Hyde Park. Walking through Hyde Park you could see couples on the ground making love to each other. In other parts, young adults played ball, and, in a special corner, there were speakers on podiums addressing the assembled crowds on topics from politics to religion.

Life at the Solomons was well-regulated and mostly uneventful. But there were two events that stand out in my mind and probably had an enormous influence on me. One was a dinner that the Solomons held for their wealthy friends. When they discussed politics, I had the nerve to tell them in my halting English about the many German tanks and airplanes I had seen on my trip to London and had the audacity to warn the guests that the Germans could overwhelm the British. I was told to sit down and my comments were taken as the fears of a little Jewish refugee boy in strange surroundings.

The other event was, on a personal level, more significant. I did not have many friends and confided to Beryl, the Solomon's son, about my interest in learning about the world. He seemed like the kind of person who had a social life, including interaction with some women, and possibly also knew how to get along with members of the opposite sex. Beryl said one day that he knew a Jewish woman, just a little older than I, who would be interested to learn about my life in Austria, and who took an interest in young refugees. A few days later, he arranged for me to visit this lady who lived on the other side of Hyde Park in one of the row houses that are typical of that part of London. The lady's flat was on the second floor. I knocked on the door and a woman in her early 20s opened it. She asked me if I was the refugee from Austria, and then motioned me to come into the flat, which consisted of just one room with a big couch and minimal cooking facilities. She asked me about my past, my parents, why I had come, and about the girlfriend I had left behind in Vienna. It was probably the first time that a person in England seemed to take a genuine interest in me. She then warmed something like cocoa to drink and we sat down together on her couch. She put her arms around me and we kissed each other. I am not sure just what language we used but she had some knowledge of Yiddish, which helped to establish contact. But, it was not only verbal contact, because she allowed me to touch her breasts and lie down next to her. I had never had sex with a woman before, and it appeared as though she would encourage me to have my first experience with her. Then she said to me, "Wait a moment, I must excuse myself," and got up and went down the hall to where the bathroom was located. While she was gone, I suddenly became afraid. I wondered what kind of woman she was. I had heard about prostitutes' sexual diseases, but my main fear came when I realized it was past the time I needed to return to the Solomons'. When she came back, I said, "I must go home because otherwise I will be punished. May I come back tomorrow?" She said, "Oh, why don't you make an excuse? You can stay with me." But I was, at that point, already unable to respond to her and said, "Please let me come back," and left. I ran from the apartment across Hyde Park to Argyle Road to where the Solomons lived. As I had feared, the front gate was locked when I arrived. Thus, I had to ring the bell and knew that my tardiness would have consequences. I did not sleep much that night and left school at noon the next day to return to Edgeware Road, where the night before I had met the woman. I knocked on the door and asked for her, but was told by the concierge that she had left. I would have probably thought that my experience of the previous night was only in my imagination, if it were not for a slip of paper I held in my hand with the address where I knew a young woman had opened the door for me.

On returning to the Solomons', I was quite shaken. I was severely reprimanded for not obeying their rules and told that I would have to leave. There was really no place to go. Somehow the Jewish organization was able to get in touch with my aunt and uncle, who had in the meantime arrived in America, and they suggested that a former acquaintance of theirs by the name of Joe Fenton might take me in. I was told to go and see Joe. He was an impresario and had met my uncle some years before in connection with arranging a contract for one of his operettas in England. He lived with his brother in a small apartment and they had an extra small room with a bed. Each of the two men weighed at least 250 lbs., and they seemed to spend their entire day preparing one gourmet meal after another. They were real gluttons, but seemed to have fallen on hard times.

The evening meal on April 20th was outstanding, but the aftermath put me in a terrible bind. The older brother told me that "Yes, you could stay with us, but you will have to pay for your keep." I said "I have no money and am not allowed to work." They then suggested that I ask my aunt to pay for my room and board. I had heard a few nights before about refugees from Vienna being shipped East, and was afraid that if my aunt had to pay for my keep, she would not help my parents escape. So, I refused to ask for her help and the Fentons said that I could only stay a few more days and should look for some other place to live.

I remembered that a former classmate of mine who had been a member of our cabal in Vienna, Rudi Bunzl, had also emmigrated to London and I asked the Jewish agency to contact the family. The Bunzls had indeed come to London and had an apartment in which they allowed me to stay while their son Rudi was away in a private boarding school. The Bunzls had been wealthy people in Vienna who had owned a chemical factory, but I do not know what they were able to salvage and bring to England.

When they took me in, they had a small apartment on the fourth floor of a building. After Rudi came back, I shared the room with him for a while but I knew that this was only a temporary arrangement and that I would have to find a way out. This was not easy, because I had no work permit and an employer in England would never think of hiring an illegal immigrant that had no work papers. Just by coincidence, however, I learned that because of a shortage of labor on some of the farms, one was permitted to volunteer as a farm worker. I submitted my name as a volunteer, and soon after received information that there was an opening—not in England, but in Wales—on a farm that was called Measmawr Hall. I had no idea where the farm was located, but enthusiastically said that yes, "I accept." On July 26th, I took a bus from London through the beautiful countryside of Wales to my new home at Measmawr Hall.

While I had stayed at the Bunzl's, the negotiations between England and the Soviet Union to form a common front against Hitler broke down and Stalin dismissed Litvinov, who was a Jew, as his foreign secretary and replaced him with Molotov. This made it clear to the world that the Russians did not intend to stand by the West in the war that now appeared to be inevitable. Furthermore, word leaked out that Germany and Russia had secretly agreed to divide Poland. This could only mean that Germany intended to invade that country.

CHAPTER 6—WALES AND THE START OF WAR, JULY 1939–APRIL 1940

The move to Wales was probably one of the most fortunate turns in my life. Measmawr Hall was jointly owned by a 23-year-old woman called Janet Owens and her brother John who was two or three years older. However, the person running the farm was their former governess, whom the siblings referred to as E. I learned that the parents of Janet and John had been killed in a rail wreck in Canada when the children were small, and that E had taken over then as their caretaker. They owned a fairly large farm, with an old sixteenth century stone building in which the family lived. There was also an artificial lake on the property in which, several centuries before, carp had been raised. The work on the farm was carried out by Janet, John, and one or two temporary hired farmhands, as well as E. It was hard work. I was taught how to milk the cows. Every morning E awakened me at 5:00 a.m. with a cup of hot tea. Then I had to go to the barn where the cows were kept. I soon learned to sit on the stool, grab on to the udders, squeeze, and forth came the milk. After milking we had breakfast, and then went to work on the farm. It was a really wonderful change in my life to know I was earning my own living. In London I had pretended not to be seen because I knew that I was living off of the generosity of other people and was not able to repay them in any way. On the farm, however, even though I was probably not a very efficient farmhand, I felt as though I was beginning to become my own person. Moreover, Janet, who had a degree in Agriculture from the University of Abrustwyth, was also interested in teaching me to speak English better, and every evening before going to bed we practiced. This was fun, since there was only candlelight in the room and we had to go to bed soon after it turned dark.

Our only entertainment was a small radio in the living room to which we could listen after we came home from work. For me, one of the most enjoyable things to do was to ride a Welsh pony that I loved from the day I saw it. I had to be careful though, because the size of the Welsh pony did not allow me to stretch out my legs. I kind of had to squash on it and lean forward to keep my feet from scraping the ground, especially when riding fast.

Life on the farm was very different from anything I had ever known. In Vienna or in London when I needed something, especially food, I would go to the market and buy the item, assuming I had the money to pay for it. But Measmawr Hall was far from a market, and although the Owens were not poor, the family prided itself on being self-sufficient. I learned that when I came to one of our first breakfasts. After the wakeup tea at 5:00 in the morning and milking the cows I was starved. On the breakfast table I found bread, butter, marmalade, milk, and porridge. And, of course E brought a steaming kettle of English breakfast tea. Unthinkingly, I first ate some porridge and then fixed myself a sandwich with a lot of butter and marmalade. After breakfast E took me aside and said "Everything on the table comes from our farm, except the marmalade. We want our life to be sustained by what we can produce here ourselves. You can eat all the butter and cheese you want, but please be sparing of the marmalade that we must buy." Although I knew that tea also had to be bought, it was apparently an exception. But the entire philosophy of the farm was to strive toward conservation and self-sufficiency. The concepts of self-sufficiency and sustainability were a basic philosophy that became a guide for my future life.

After a few days, I learned how to harness a horse and to harrow a field. That is, I let the horse pull a huge rake that did something for the field that was apparently necessary. We also spent a lot of time harvesting turnips. That meant you would grab a turnip that was in the ground, pull it out with your left hand, then use a hatchet in your right hand to cut off the green stuff and throw the turnips into a wagon that was following alongside of those of us who were harvesting the turnips.

A particularly memorable day was Rosh Hashanah, when I decided to fast, but still go to work. I was very much aware of the festivity of the day and how thankful I was that I could be giving a kind of prayer in the setting of a beautiful field in a place where people were friendly and did not look at me with hate or generous pity.

A week or two after I arrived, another boy came to help on the farm. His name was Ernst Phillips and he was approximately my age. We immediately became very close friends and formed a bond that lasted throughout my stay on the farm and beyond. But one day we did a very foolish thing. Ernest and I decided that we wanted to find a job that paid actual money and took off on our bicycles in search of employment. We did indeed find a farmer who offered us a job on his farm, mostly dealing with taking care of his sheep and cows. After Ernst and I arrived on our bicycles, the farmer asked me whether or not I knew how to milk cows. I had performed this task at Measmawr where the farm had four milk cows. E had shown me how to hold the udder, close first my index finger and then squeeze the rest of my hand so that the milk would squirt into a bucket. I thought I knew all about how to milk cows and told the farmer yes, of course, I knew how. Then the farmer took me into the cowshed and showed me a mechanical attachment made of rubber that could be attached to the udders and, by means of a vacuum, pull out the milk. The procedure seemed quite simple and the next morning I proceeded to milk the 20 cows owned by the farmer. Everything seemed to be going well, when after about five or six cows the farmer's wife came running into the shed screaming, "You are running the milk over the buckets into which you are to catch it." I realized then that the tube from the cows ran into one of six buckets that were aligned and needed to be shifted after a bucket was filled. As elementary as that seemed, I did not realize that I needed to do that and, after being dressed down summarily, both Ernst and I were told to leave the farm. Thus ended our first venture into independence and we made our way back to Measmawr with our tails between our legs. Janet and John found our misadventures hilarious and, without hesitation, took us back into the fold of the Owens family. They did however have a good-natured laugh at our expense and merely told us to "Go and have some milk."

While we were working peacefully on the farm, war clouds began to rise in Europe. I still remember the day, September 3, 1939, when we sat by the radio and Prime Minister Chamberlain declared that a state of war now existed between Germany and England because Hitler had unleashed his Panzers and invaded Poland. We followed the events after the invasion of Poland by listening every night to the radio. It did not take many days before Warsaw fell and Poland capitulated. Afterwards, Russia and Germany divided Poland in accordance with a secret agreement of the German–Soviet peace pact that was negotiated a few days before the war began. Not very much changed in our life at the farm in those early days, but we heard about people being inducted into the British Army and John left to join the RAF. I did not see John again, but many years later learned that during the Battle of Britain, when, as Churchill put it, the "world at large owed an enormous debt to the few that fell," John had flown one of the Spitfires that defended England during that battle and had crashed on the farm after his plane was hit. He died there and was buried at Measmawr Hall.

One day in October I received a notice from the British government that I had been declared an enemy alien and was to be interned. This was the fate of the majority of the male refugees in England at the time. But thanks to the efforts of Janet and E, I was given an opportunity to explain to a special tribunal that, although I was born in Austria which was now part of Germany, I was Jewish and came as a refugee in a children's transport. Hence, I told the tribunal "I love this country and am not an enemy of England." I was very lucky, because the majority of refugee boys that had come to England and were over 16 years old were declared enemy aliens and were shipped off to Australia and interned. I was allowed to stay on because the investigation clearly established that I was in no way a threat to the country.

Afterwards, however, I decided that I wanted to participate more actively in fighting the Germans, only I did not know how. First I volunteered for the RAF and not unexpectedly was rejected. But in the train station on my way back from the tribunal I had helped an elderly woman carry her suitcase. I told her about my situation, and when she thanked me for helping her, she asked for my address. A few days after I returned to the farm, I received a call from the woman. She told me that her son owned a factory that made parts for Britain's Glenheim Bombers and probably could use some help. That factory was located in Hereford, not too far from Measmawr Hall. Janet understood my desire to contribute to the war effort and encouraged me to apply for the job. It was a tearful good-bye and I took off with my few belongings in a rucksack, including the tin box that I mentioned previously. That tin box now contained some money that my Uncle Siegfried had sent me from Shanghai. I believe it was 20 pounds, which to me seemed like an enormous fortune. In addition, I had been instructed by my father to keep all of the envelopes of the letters he sent to England. He had collected stamps in Austria and, since he could not send out any money, he thought that one way to help me would be to put special commemorative stamps on the letter envelopes. Sometimes he also included some extra ones in the mail. The German censors, unless they opened the envelopes, could not distinguish between paper on which writing was done and stamps, which were also paper. It is with the help of the postmarks on those letters that I have been able to track my moves during the first tumultuous six months of my immigration to England.

When I arrived by train in Hereford, I immediately began looking for a place to stay that would be affordable on the wages I expected. I could not find a place of my own, but a woman told me that I could share a room with another fellow of approximately my same age who was a day laborer from another city. I was a stranger who had no means of real communication and therefore reluctantly accepted this lodging. It was a great mistake.

When I returned to my lodging the evening of the day I started my job in the aircraft factory, I found to my horror that the tin box which I had brought with me had disappeared, as well as the fellow who shared the lodgings. I was desolate. Not only was my entire world's fortune in that box, but also the Omega watch that my father had given me as a parting present. I went to the constabulary and informed them of the theft. They were able to track down the thief. He was a criminal with a previous record and the police arrested him the next day. He was charged with robbery and theft, and a few days later, I was asked to go to court where this man was brought to trial. He admitted to having stolen the tin box and returned it with the envelopes. However, he said that he had used the money to pay off debts at a tobacconist and for a lottery where he was betting on horses. Also, the watch had disappeared. I do not know what happened to the man. He may have been sentenced to jail, but as far as I was concerned it didn't matter. Everything I owned in this world, except for the clothes on my back, had been stolen.

I continued working at the factory and living in the boarding house. The working hours at the factory were soon increased to between 70 and 80 hours per week. This increased my paycheck to 7 and later to 8 pounds per week, an enormous sum in my eyes. There were three other unmarried men from the factory living in the boarding house. All four of us ate at the same table and talked about the war, our work, and the condition of labor. All three of the men belonged to the Amalgamated Engineers Union that everyone had to join in order to work in the factory. I was classified as an apprentice machinist. The men explained to me how the union had improved working conditions and wages over the years. They were all strongly anti-German and anti-Russian, but also in favor of Socialism and often talked about the benefits England would derive from a Labor Government after the war. When I was told that I had to join this union I had no hesitation and was proud to receive my union card.

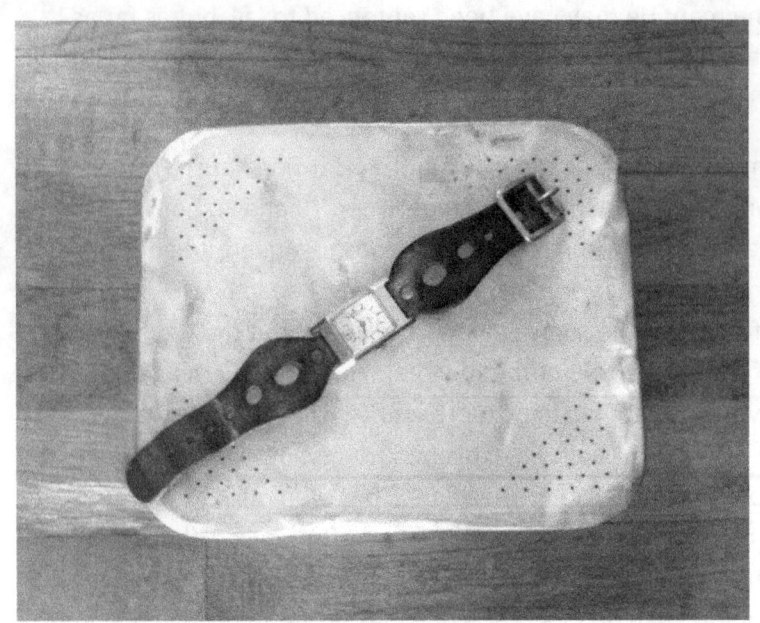

I want to deviate from the sequence of this book, because the topic of the watch which was stolen from me in Hereford had a very curious ending. Before leaving England to go to America, I left with the local Hereford police a description of the watch and information about my forthcoming emigration to the United States. I also told the police that my parents lived in New York and that this was my immediate destination. Soon after arriving in New York, as I shall recount in more detail later in my story, I realized that New York was not a place where I could find a decent job and where I wanted to remain. When my friend Paul Toch invited me to come to California, I wasted no time and left New York by bus on a long journey to join him there. Several years later, when I was still working nights in a machine shop, I received a letter one day from the Jewish Agency in New York with a clipping from the New York Times. The clipping was an article entitled "Watch Looking for Man Around the World." It recounted briefly the story of a young refugee who had been robbed of all of his belongings, including an Omega watch, and that the watch had been retrieved by the constabulary in Ireland. The information that was left with the police concerning the boy having worked in the aerospace industry was probably referring to my employment with Aeroparts in Hereford. The Jewish Agency in New York had kept track of my whereabouts and connected the story I told them of the robbery in Hereford with the story about the watch. I replied, giving them my current address, and asked that the agency inform the constabulary of my whereabouts. Another few weeks elapsed; then one morning, quite unexpectedly while I was still asleep, the doorbell rang and an FBI agent asked for Franz Kreith. To my amazement and surprise, he then asked me to describe the watch I lost in Hereford. He compared my description with the watch that he had brought along, and since the description matched, he handed me the Omega watch which my father had given me in Vienna many years before. I subsequently tried to find out how the watch was retrieved in Ireland but had no success. The watch was very dear to me, but the watchband had deteriorated badly. I had a cobbler make me a leather watchband and then used the Omega watch for many years, until I replaced it with one that had date, time, and alarm functions. However, I still have the tin box and the watch in a drawer next to my

bedside and every so often look at them with amazement. They remind me that the watch, the clothes on my back, and the box were all of my worldly belongings at the start of my life.

I had gotten the job at the factory by pretending I knew more about machinery from my brief schooling at the Polytechnic Institute than I really did. I had learned how to read blueprints and work with sheet metal, but in the factory I was asked to work on drill presses and milling machines. I was a fast learner, however, and the workers around me were kind and taught me how to do the work I was expected to do. I still remember one particular job. It was on a milling machine and one had to cut slots into a ring which was a so-called spanner used to tighten the propellers on the Lancaster bombers. It took approximately five to six minutes for the milling machine to make each cut. The work was extremely monotonous, but I managed to use the time between cuts to improve my English. In the evening I would read a book from the library, write down all the new words that appeared, and translate them from a dictionary. I then wrote about 20 or 25 words on a piece of paper and, between each cut, I pulled out the paper and tried to learn one or two new words. I repeated these words during subsequent cuts until they became firmly embedded in my mind so that I could use them. It was during that period that I again felt very useful because I was helping the war effort. At the same time, it was a blessing because I improved my English vocabulary every day by at least 10 or 20 words.

The great event of that winter was a visit to my little sister, Susi, who had been taken in on a later children's transport by a doctor's family in Brighton, near the coast of England. I had earned enough money to pay for a rail ticket to visit Susi, whom I loved more than anything in the world. She had become a sweet little English girl, spoke fluent English, and the family loved her. I only owned one pair of shoes at that time, and as luck would have it, the sole of my left shoe came off unexpectedly on the train trip to Brighton. It had snowed and it was very slushy outside; with every step I took brought some of that slushy snow into my shoes. It was Saturday afternoon when I arrived and two days before Christmas. I frantically tried to find a shoemaker who could sew my shoes together, but of course, in a new city, not being able to walk very far or very fast, I had no success. The only thing I could do was to take off one of the shoelaces and tie it around the sole of the shoe as best as I could and then go back to Susi and her guardians. They wanted to take a walk, but I was terribly embarrassed and declined, of course. I limped around for the rest of the weekend and was fortunately able to find a shoemaker on Monday who could stitch the sole to the shoe. This sort of thing is very hard to imagine today and it is, therefore, not surprising that the image of a left shoe with half of its sole disconnected is still with me. But, Susi was happy to see me and I was, of course, delighted to find her in a home where the family was so good to her.

I remember that Christmas in Brighton as very beautiful. Many years later I learned that Doctor Maysell, who had taken Susi in, had died during an air raid. He had an appendectomy the day before the air raid's sirens came on and tried to get to the air raid shelter when his stitches came undone. He died from the aftereffects of the bleeding that ensued.

Upon my return to Hereford after my trip to Brighton, I found no news from my family in Vienna. As a result of the war, it was impossible to send letters directly to family members in other countries and the Red Cross had recently become the intermediary for communications. I continued working at the aircraft factory and soon became part of the unionized working force. I also learned a good deal about the attitude of labor towards management. But, at the same time, I saw the engineer in a white shirt sitting behind a glass door and began to think that it would be nice to be an engineer instead of a machinist. I became a pretty good machinist in the six months (November '39 to April '40) that I was at the Aeroparts Machines Corporation, and learned a good deal about lathes, drilling presses, and milling machines. This was very fortunate because it helped me and my family to earn a living after I came to America.

CHAPTER 7—GOING TO AMERICA (APRIL 1940)

At the end of April, 1940, the Jewish agency informed me that visas for me and Susi had been obtained and I was asked to come to London in order to prepare for the trip to America. However, because ships between England and America were continuously attacked by German U-boats, ships had to wait for a convoy to assemble. It was therefore not possible to set a specific departure date. While waiting for an opportunity to find a boat, I was put up in a rooming house somewhere in the seedy part of London—I don't exactly remember where. Susi remained with her adoptive parents in Brighton because it was only a short distance from Brighton to London and she could join me on short notice.

The boarding house was an old dilapidated place that had been converted into a holding station for refugees waiting to emmigrate to many different parts of the world. All of the people milling to and fro in the entrance hall had one thing in common: They were all Jews who had fled from one of the countries on the European Continent to England, but were not allowed to stay in England. They had to move on, but only a few other countries would accept Jews. The exceptions were Cuba, Shanghai, and a few countries in South America and Australia. Only a handful of the people had affidavits to go to the United States. Although we were all Jews, the place seemed like a Tower of Babel with people speaking at least six or seven different languages, searching for friends, and ceaselessly asking about news, since many could not read English newspapers. There was a radio in the entrance hall and a crowd always stood around it listening to the BBC. It was one of the darkest hours of the war.

With England at war, London was dark at night and the rooming house was a dismal place. Food was short and people were uneasy because they did not know what was in store for them. Unexpectedly, however, one of the most important and significant events in my life occurred in this Kafka-like setting. I met and fell in love with Herta, a young woman my age in the rooming house waiting to go to Shanghai. I believe that it was about 13 or 14 days after we met that Herta learned from one of the Jewish caseworkers that a ship would be leaving for China in a day or two and that her parents would be coming to join her. I don't believe that we ever had a chance to properly say good-bye, because all the refugees had to keep their trunks ready for departure and a caseworker could come any time and say, "Get ready, your ship has arrived." It happened one morning, and in the afternoon Herta was gone. I remained in the dismal rooming house for quite a few more days waiting for a ship to America. About a week after Herta had left, I received a postcard from her in which she said, "Until today, I thought I was carrying your baby, but not for a moment did I regret our time together. Now I know that I am not pregnant and that our time together will soon only be a memory. Good-bye." This was the last I heard from Herta. I could not answer her since I did not know where she would be in China.

I felt lonely and depressed after Herta left, but fortunately, two or three days after receiving her postcard, I too was asked by a caseworker to get ready because a ship would be leaving from Liverpool the next day. My little sister Susi arrived that afternoon with a tiny suitcase and a teddy bear. The following day we took the train to Liverpool and embarked on our trip to America. I had no photograph or any kind of memorabilia to take with me as a physical reminder of Herta, but the experience must have been very profound, because it was many years before I again touched another woman.

Susi and I embarked on a merchant ship, the S.S. Antonio, which had room for a few passengers. The ship was part of a convoy protected by an old battleship, the HMS Renown. She was a huge ship, built originally in 1916. Later in the war she took part in the hunt for the German battleship Bismarck and carried Winston Churchill to and from a Teheran conference with President Roosevelt and the Russian Dictator Stalin in 1942. She is one of four battleships that survived the war. I am still amazed how calm I was, because we knew that a trip between Liverpool and New York would be dangerous with German U-boats lurking in the wait to torpedo our ship. Moreover, while we were crossing the Atlantic we learned that Hitler's army was invading Holland and France. But the overriding feeling for both Susi and me was that we would soon be reunited with our parents. Part of this expectancy was a feeling that my days of having to fend for myself would be over and that my father, whom I remembered as a strong and decisive person, would take care of me and Susi from here on.

The trip was long, but fairly uneventful. I remember that the ship at times would take zigzag evasive actions which presumably were necessary to evade the German U-boats. But a few days after we started our journey the ship's radio brought us some terrible news. After Hitler had conquered Poland and divided the country with Russia the Germans took no further military action against the Allies. England had sent a military force to France and both the French and the British Armies sat behind the Maginot line, a defensive barrier that the French thought to be impregnable. This period of seven months was a lull in military activities, often referred to as the so-called "Phony War." But the stalemate ended suddenly when the German Army went around the Maginot line, invaded the Netherlands on May 10th and quickly subdued the country. Three days later, the German army burst through the Ardennes and on May 20th reached the Channel Coast, thereby isolating the British Expeditionary Force from the French Army. The British were forced to evacuate about 340,00 soldiers from Dunkirk between May 26 and June 4 by a hastily assembled flotilla of 700 merchant marine boats, fishing boats, pleasure craft, and RNLI lifeboats. The operation was called the "miracle of small ships." The evacuation was successful in that it saved the major part of the British troops, but the Allies left behind their guns, ammunition, trucks, and supplies. We listened to this disastrous turn of events on the ship radio on our trip to the U.S. and expected the Germans to invade and conquer England any day. The reason Hitler decided not to cross the Channel is still not clear. Some historians believe that he expected the British to join the Germans in the planned invasion of Soviet Russia. We will probably never know for sure, but for the Jewish refugees doomsday seemed to be close at hand.

Then, three or four days before our scheduled arrival in New York, we were told that the ship would not go to New York, but had been ordered to proceed to Halifax in Canada instead. From that time on, we were quite anxious, but the trip continued to be uneventful. Two days later we actually arrived in North America. We were met at the dock by someone from the Jewish agency and quickly put on a bus to New York. We crossed the border at Rouses Point on June 12, 1940 and arrived in New York a few hours later. When we stepped out of the bus we saw Father, Mother, and my younger brother Kurt waiting for us. It was a moment that, for a long time, I had waited for and had great expectations about, even though I did not know if it would ever happen. However, seeing them brought mixed emotions: We were filled with happiness at the reunification of our family, but I was shocked at the appearance of my parents. Their dress was shoddy. My father, the once proud and erect man I remembered from my youth coming home on Friday nights to the "Sommerfrische" in Tyrol to bring us candy, was now stooped. And my mother, who had already deteriorated in appearance in Vienna, had gained a lot of weight and was dressed like a washerwoman. We embraced and kissed, but somewhere deep down I was aware that my hope of being cared for by my parents had vanished. Although I was only 17, I realized that from that time on, I would be the head of the family. This feeling was reinforced when we arrived at the dismal place where my parents stayed on 87th Street near Central Park. It was a brownstone home in what was at that time a poor white, largely Irish and Italian neighborhood of New York. My parents had been assigned by the Jewish agency to a single room where they slept, cooked and lived. The room was above the apartment of the concierges, who were an Irish couple by the name of Cosgrove. There were only two beds in the room—one for my parents and one where my brother Kurt had slept, and which now also had to accommodate Susi. For me, the only place to sleep was the bathtub. It was not really that bad, but after having had a bed of my own in England, it was quite a shock.

I soon realized that my parents had neither a job nor an income, and were totally dependent on the Jewish Relief Agency in New York. It was therefore only a day or two before I began pounding the pavement trying to find a job that would earn money. I did find a job at a small shop, but it was a pretty terrible way to earn a barely living wage of 30 cents an hour. The shop made small broaches for people using long spools of square wire to form various names or titles. I was shown how to bend the wire into the word "Mother" and was asked to repeat the task of bending "Mothers" out of copper wire all day long. The job was not only monotonous, but the square wire cut deeply into my fingers when I held it in place for the pliers to bend it into shape. After about eight or nine days, my hands were so bloody that I had to quit. I wished I had kept one of the "Mother" broaches as a memento of that dismal start of my life in America.

My mother, despite all of her shortcomings, had a high regard for education and agitated that I should continue my high school education to get a diploma. She managed to find a small high school in the black part of New York, called Harem High School, that was willing to enroll me, although I had no documentation of any previous education. At the same time, I continued looking for a job and finally found one at Linotype Corporation. I worked there part-time on a metal grinder, on a pay-by-the-piece basis, and went to school in my spare time. But there was one good side to my going to Harem High. The school had a soccer team and over the weekends I could play soccer again. Soccer had been my favorite sport in Vienna. I played left defensive back because I was not fast enough for left forward which was my favorite position because you were more likely to score a goal from there. In Vienna, I not only played soccer, but Ope took me to professional games on Saturday afternoons. My father and I did not have many opportunities to be together, but going to these games gave me a chance to be alone with my father. Although there was a Jewish sport club called Haakon in Vienna, we rooted for another club, called Rapid. This must be a fairly common name for soccer clubs because there is one by the name of Rapids here in Denver. I have often thought of becoming one of its fans, but it would not have been the same as in Vienna with my father and I never did.

Although I was not an enthusiastic student, I did attend classes faithfully and even though my English grammar left much to be desired, I tried to hand in my homework on time and this seemed to be unusual in the school. To my amazement, after a few weeks of attending school I received a certificate which stated that I had graduated from high school. To be honest, I remember very little about what I learned.

CHAPTER 8—LOS ANGELES AND START OF THE WAR, 1940–1943

I soon realized that New York was still in the Depression and that I had to go somewhere else to make a living. Fortunately, my good friend from Vienna, Paul Toch, who had previously immigrated to Los Angeles with his family, invited me to join him. I scraped together all the dollars I could find and got a few more from the Jewish Agency to buy a one-way bus ticket from New York to Los Angeles. It was a long hot ride, because in those days there was no air conditioning and the bus had to pass through the Mohave Desert at night in order to avoid passengers dropping from heat stroke. When I arrived in Los Angeles, I was met by Paul and his family, who had, in the meantime, started to make a living. Although in Vienna the Tochs' had owned a poultry shop, his mother had always been an excellent cook. She used that skill to obtain a position as a cook in one of the fanciest hotels in Hollywood. Thus she not only earned hard cash, but was also able to bring home virtually all of the food for the family that now also included me. That was the beginning of a new phase in my life in America.

After arriving in Los Angeles, I contacted my Aunt Ellen and Uncle Heinz Reichert who had in the meantime bought a lovely house in Hollywood and seemed to have happily settled there. I asked Ellen to help my parents and my siblings, Susi and Kurt, to also come to Los Angeles, which seemed a lot nicer than New York. However, my aunt disliked my mother intensely and refused. In fact, I was later told that she had refused to pay for a ship ticket to help her escape from Europe at the time my father and younger brother were ready to leave. Fortunately my father had been able to obtain, through some of his old Viennese friends, enough dollars to also buy a ticket for my mother and so she too was able to come with him to America. But now, my aunt refused to have my mother live near her, and only father and Kurt were invited to come to Los Angeles. I am not sure how this sad situation was resolved. But some time after my father and brother arrived in California, while my mother remained in New York with Susi, father was somehow able to convince my aunt to let my mother and Susi join him. Soon after the family was once again together we rented a place on La

Cienega Blvd. and all three of us adults immediately began to look for work. But we were poor and life was hard. I can still remember that my father who always "liked something sweet " at the end of dinner went every morning to the bakery down the street to look for "day-old" leftovers because we could not afford fresh bakery. He also bought himself shoes at the Salvation Army that did not fit properly and gave him blisters or corns; but he never complained. My mother suffered from a more severe affliction. She liked to go shopping, as she was used to in Vienna, especially to buy presents for other people. However, because she did not have enough money to buy any of these presents, she made a down payment on them. But since she could not afford to pay for the rest she owed, the down payment was lost. Moreover, she rarely told us about these "purchases" and after my father found out about these idiosyncrasies, he had to keep all hard money hidden from his wife, which she of course greatly resented.

Eventually my mother found a part-time job as a night nurse and my father was able to get occasional work as a gardener. I found a job in a machine shop which specialized in stamping sheet metal parts from large strips of metal for other organizations. The place was owned by a middle-aged man named Emil Schigut. At the time I took the job I noticed that two fingers on his right hand were missing. When I asked him what happened, he said "I had an accident and one of the stamping machines took off my fingers while I was pushing the sheet metal strip towards the stamper." That story scared me, because I noticed that some of the stamping machines were without the protective shields that would avoid such an accident. When I asked Mr. Schigut why the machines were without such protection, he said that these protective devices were a nuisance and reduced the speed at which a worker could do his job. He then cautioned me to be careful.

Schigut Stamping was located in Culver City, and in those days there were still streetcars going from Hollywood, where we lived, to Culver City. Hence, I could commute to work via streetcar and did not need a car. Although I felt apprehensive about the conditions at the small factory, it was a job, and I continued my work at Schigut for several months. My interaction with the other workers was minimal because they were so very different from me. I do not remember any of them in particular, only Mr. and Mrs. Schigut.

Because I felt so uncomfortable at the Schigut factory, I soon started looking for other job opportunities. That was not too easy, because I could not take time off from work and had to do my job search on Saturday mornings relying only on ads in the newspapers and word-of-mouth information. Finally, I did find a job in a machine shop which used the kinds of machinery that I had learned to operate in Hereford— such as drill presses, milling machines, and lathes. Of course, my skill level was still somewhat limited and I had oversold myself as a machinist. As a result, I did not last very long on this or any other of my first few jobs. But in each of them I picked up some additional skills and continued to improve my abilities as a machinist. After about six to eight months, I was able to handle almost any job on a lathe, milling machine, or drill press and could call myself a skilled machinist.

But working in the American machine shops was very different and less pleasant than in England. In Hereford all of the workers belonged to a labor union and my co-workers treated me like an apprentice. They were always willing to help and give me instructions when needed. In the American environment, however, I was more like a competitor, someone who had taken a job from one of their people. Many of my co-workers had come from Oklahoma after a period of drought and sandstorms destroyed their farms. They were called Oakies. A few others were from the Middle East and clearly disliked Jews. They teased me, asking if I was Jewish. To my everlasting dismay, I did not proudly say YES, but evaded the question and said "I am not religious." They of course knew that I was a refugee from Europe and in their eyes continued to be "a yid." This became painfully obvious one night when one of the Arabs took a white hot metal filing from a lathe and stuck it into the back pocket of my pants. He laughed when I winced with pain.

My friend Paul and I managed on many occasions to sneak off to the mountains for hiking and skiing. The most memorable of these trips occurred in the early part of December, 1941. Paul, a girl friend of ours, and I went on a ski trip in the mountains near Los Angeles. After a day of hiking on the skis, we found an abandoned hut in which we stayed overnight. The following day we further continued into the mountains and towards the evening we saw a light in the distance. We skied towards the light and found a small farmhouse. The farmer offered us lodging and an evening meal. It was a cozy group and we thoroughly enjoyed the experience, but when we awoke the next day, the farmer told us that the United States was at war. It was December 7th. The Japanese had bombed Pearl Harbor while we were cozy in our sleeping bags and the United States had joined the war against Hitler.

I continued my work as a machinist at Johnson Tool and Die in LA, but the orders for the shop came now mostly from the military. One of these orders was from the U.S. Navy for water distillation units on lifeboats, and I was asked to apply for security clearance. I filled out the necessary form and received a temporary permission to work on war-related orders in the machine shop. I also received a temporary deferment as the head of a household working as a skilled machinist on war contracts. My feelings were of gratitude to be able to assist in the war effort, despite the fact that I was not yet an American citizen.

I tried to avoid fraternizing with the other workers. They often went out for a beer after work, while I tried to go home quickly and read or play tennis. This pattern became imperative for me later on because I decided to continue school. My decision to continue with school changed my life. I can still remember vividly how this decision came about. One morning, I was playing tennis at one of the public courts where people had to wait until an opening occurred and then jump in. One of the players waiting with me was a young woman who told me that she was working at night and going to school in the daytime. I said "Gee that is interesting." She replied "Yep. But you have to be very tough. I wonder if you could do it." I replied "I can do that too," and she challenged me to try. We made a date to play tennis together again two months from that day and I would report to her my progress. As it turned out, I took the bait and applied for admission at Los Angeles City College (LACC) a few days later; thus started my college career. Unfortunately, I never saw the young lady again and was unable to tell her that I had met the challenge that she initiated at our tennis game.

Of course, in order to go to school during the day I had to find a night job. This was not too difficult, because the war was raging in Europe and, although the United States was still officially neutral, many of the machine shops had extra orders and were beginning to put on a swing shift that worked from 4 to 12 and a graveyard shift that went from 12 to 7. I worked on both of these shifts in various shops and managed to go to school in the morning. The first year I had little time to study and barely passed my elementary physics and math courses. I remember that my grades were poor and I received a barely passing grade of D in Calculus. I don't think I really learned very much during this period of work and school, because most of the time I was tired and hardly able to handle both of my lives—the night life at work and the daytime at school. However, for the third semester I took some easy courses, including one in German literature and after a year and a half managed to do well enough at LACC to gain entrance to the University of California at Los Angeles (UCLA) in the fall of 1943.

My family's financial situation was still quite precarious, but had improved some with my parents' part-time employment. However, their income was inadequate to maintain a family of five. My deferment offered me the opportunity of continuing my schooling at UCLA and actually completing my sophomore year on a part-time basis in 1944. My wages were fairly good, and combined with the earnings from my father's and mother's jobs, we were able to take the first steps from being dependent on welfare to becoming able to care financially for our family. In fact, we were able to make a down payment on the house on 722 Westbourne Drive to which we had moved, and thus began our path towards the American Dream that for us was financial independence.

After my six month deferment to work in the war industry expired, I requested another deferment. However, the draft board informed me that I could no longer qualify as head-of-household and would therefore have to report for induction. I appeared at the military induction center in Los Angeles and was examined for fitness to serve. Quite frankly, to my big surprise, the doctor told me that I had bad asthma and a weak heart and would therefore be disqualified for active service. To this day I do not know what the symptoms were that prompted his decision, but once I knew that I could not serve in the American Army, I decided that I would try to complete my engineering education. At that time, there was no engineering college at the University of California in Los Angeles, and in order to continue my education, I had to transfer to the University of California at Berkeley. I was only four semesters, or roughly sixteen months, away from graduation and was therefore able to borrow money to pay for tuition as well as for room and board at Berkeley. In addition, with my skills as a machinist, I was occasionally able to find work during vacations, even if it was only for a week or two.

CHAPTER 9—BERKELEY, FEBRUARY, 1944–JULY, 1945

The cheapest place to live in Berkeley at that time was the co-op on Oxford Street. It was an old dilapidated two-story house in which arrangements had been made for barracks-style living for approximately 30 students. The house had a large kitchen and food was prepared not only for the 30 students living at that co-op, but also for other ancillary places that served as extensions of the co-op. It was not a great place, but there was an *espirit de corps* that kept us going, and I managed to complete my Bachelor of Science in mechanical engineering at Berkeley after the first summer semester in July of 1945.

I was not a particularly good student, but there was at least one topic that attracted my interest—heat transfer. At that time, there were only two places in the

United States where this particular subject was taught, MIT and Berkeley. The person who taught heat transfer at Berkeley was Professor L.M.K. Boelter. Professor Boelter had a small group of students who were devoted to him and became known as the Boelter Boys. I was not officially one of them, but managed to take courses in heat transfer with two of Boelter's assistants, Ray Martinelli and Earl Morrin. The courses fascinated me and, although my grade point average was only a B minus, in heat transfer I managed to obtain an A and also do some independent work which brought me to the attention of Boelter and his cohorts.

The political scene at Berkeley in general and the co-op in particular was clearly left wing. There were several left wing organizations and clubs on campus, some of which were subsequently declared Communist fronts. I was pretty naïve politically, but considered myself a Socialist and, in particular, continued my belief in the importance of unions as protectors of the working class. This belief in unions had started at the aircraft factory in Hereford, and I still believe that despite some of their shortcomings, unions are an important political force for the good of average workers. For entertainment, I went to movies, lectures and dances, many of which were organized by the American Youth for Democracy. At one of these dances I met a young woman called Doris Monford. There was a strange attraction between us, but we were not a good match. She told me that her father had been an organizer for the Longshoremen's Union under Harry Bridges in Seattle and that she was a member of the Communist Party. That did not bother me particularly, although she tried on numerous occasions to have me also join the Party. But, I was not a joiner and was far more interested in finishing school than becoming involved politically.

CHAPTER 10—JET PROPULSION LAB OF CAL TECH, LOS ANGELES, 1945–1949

After graduating from the University of California at Berkeley during the summer of 1945, I could hardly wait to get back to Los Angeles where my parents and siblings were still living. The afternoon of the day of my last final exam, I packed some clothes and hitchhiked to Los Angeles, where I arrived late that night. My father had borrowed some of the money for my last semester at Berkeley so that I would not have to work part-time and our financial situation was critical. I therefore started looking for a job the day after arriving in Los Angeles. Almost simultaneously with my graduation, Professor Boelter was offered the position of Dean of the College of Engineering at the University of California at Los Angeles. My first effort, therefore, was to go and see Dean Boelter and ask whether he could use me at UCLA in some capacity.

FOUNDING DEAN

Llewellyn M. K. Boelter

Professor Boelter was not in his office, but on his desk was a letter from the Jet Propulsion Laboratory (JPL) of Cal Tech. It lay in a slanted position and I sneaked a couple of glances, although I knew that this was not the right thing to do. From the words that I could see, I gathered that JPL offered Professor Boelter a position to head up heat transfer research at the newly formed laboratory. When Professor Boelter returned, he told me that there was no opportunity for me at UCLA, but suggested that I try JPL. After returning home, I immediately phoned the laboratory and was told to come for an interview the next day. I went with great trepidation, remembering that the job in heat transfer had been offered to an esteemed professor. But at that time I had unlimited confidence in myself. So early that morning I drove to Pasadena and found my way to the JPL, where I was met at the entrance by Dr. Frank Malina, who had an intermediate level administrative job at the newly formed lab. Dr. Malina and I must have hit it off because he offered me a position on the spot.

I did not realize at the time what a momentous opportunity my work at Cal Tech would be. I knew that the Germans had fired advanced V-1 and V-2 rockets at London towards the end of the war, and that obviously the United States was trying to learn more about rocketry. I did not know, however, that this was the beginning of man's effort to go into space. I learned later on that Frank Malina had written a PhD thesis in which he developed the criteria necessary for rockets to escape from the gravitational pull of the earth and that his analysis would become a cornerstone of our efforts to design and build rockets to reach the moon a few years later. I met Frank again many years later in Paris where he had become a famous artist. One of his intricate mobiles was exhibited at the Louvre, the great French art museum. Frank had left the U.S. when the House Un-American Activities Committee began its Communist witch hunt and requested that he appear before the committee.

The director of the laboratory was Theodore Von Karman, a Jewish refugee from Hungary, who was an international figure in heat transfer and fluid mechanics, as well as in rocketry. I knew his name from my readings in heat transfer, but was not prepared for the imperial presence of the man who presided on Monday mornings over a conference at which we had to present monthly progress reports.

Soon after I arrived at JPL, Von Karman moved to Washington in order to assist in the development of rocketry and help with the war effort. His contributions to the subsequent achievements of the United States in the space race were acknowledged in 1994, when the U.S. Post Office prepared a commemorative stamp in his honor. The copy of an envelope above shows Von Karman on the right and various key events in the country's effort to land a man on the moon.

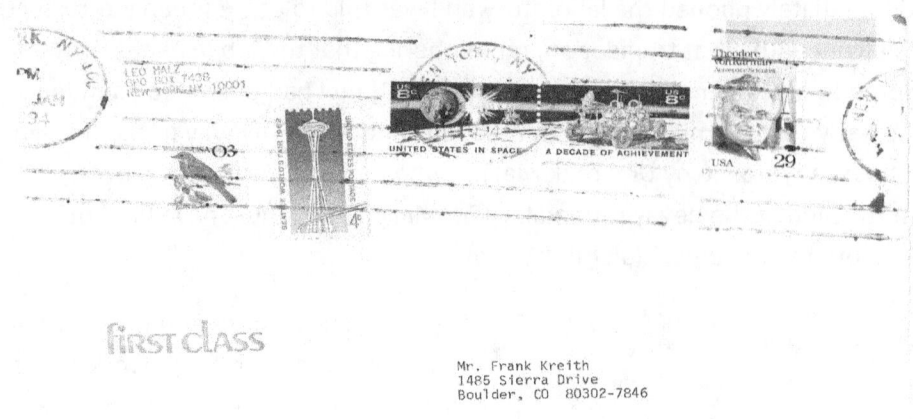

Mr. Frank Kreith
1485 Sierra Drive
Boulder, CO 80302-7846

The Jet Propulsion Laboratory was almost an hour's drive from my home on Westbourne Drive in Hollywood near La Cienega and Hollywood Boulevard. To save money, we formed a carpool of four people living in my vicinity for the rides to and from work. Although the commute took two hours of each day, time passed quickly in conversation and exchange of information about rockets.

One of our riders, Dr. Bob Boden, turned out to be a real racist who hated Blacks with a vengeance. Bob had played football at Georgia Tech and sustained a lower back injury which partially paralyzed his lower spine. He was a friendly, outgoing person, except when the topic of Negroes came up. He was married to a slightly built woman whom he called his "strawberry blonde" and whom he adored. He was constantly afraid that a Black man might try to make advances to his wife, because he believed that the penis of a Black man was considerably larger than that of a White man, and that once a woman had sexual intercourse with a Black man she would never be satisfied with a White man. This may have been an inferiority complex on his part, but he had additional vague ideas about Blacks which were inconsistent with his education of a PhD from Georgia Tech. He believed that all Blacks smelled bad and excelled in sports because they were closer to animals than Whites. There was no sense reasoning with Bob on these topics—he came from the South and just "knew more." Although I soon learned to avoid the topic of race, my association with Bob was an indirect window into the mind set of racial hatred. I also realized that there were bigots in the U.S., and racial prejudice was not restricted to the German and Austrian Jew-haters that I had personally experienced on an institutional basis. The second of our colleagues was a Jewish guy from Brooklyn by the name of Devorkin. He was short, stocky, and spoke with a Brooklyn accent. I don't know how Bob Boden felt about Jews, but he never made an anti-Semitic remark in our presence. The fourth man in our company was not "a steady" and had no particular impact on my life.

Soon after I graduated and began working at the Jet Propulsion Laboratory, I met J. Robert Oppenheimer who had returned to Cal Tech after having been the Director of the Los Alamos Nuclear facility that developed the atomic bomb that was dropped on Hiroshima. Although we all know that this bomb brought the war with Japan to an end, there are many who question the necessity of having used a nuclear weapon. Robert J. Oppenheimer must have had some misgivings about his participation in the development of this weapon of mass destruction because he remarked later that his success brought to his mind the words from the Bhagavad Gita: "Now I am become Death, the destroyer of worlds."

At that time, Oppy, as he was called, sponsored a group of young engineers in an effort to provide information to the public about the peaceful uses of the atom. We were given brief background lectures on the atomic bomb, nuclear fission, nuclear power, and opportunities for using nuclear energy for peaceful purposes. I recall going to several of these meetings, which were designed to prepare us for giving talks to the public about nuclear energy at a very elementary level. Oppy was present and gave a brief presentation at one or two of the meetings. I must have been very self-confident then, because based upon this brief introduction I presented several lectures at church groups and youth organizations. But since I had no direct involvement with nuclear energy in my job, I became much more interested in working on rocket motors at the Jet Propulsion Laboratory. However, I retained a certain hope that nuclear fission could be used for the betterment of mankind. From my basic background in thermodynamics I knew that nuclear fission could generate a large amount of heat and that this heat could be used to make steam to run a conventional power plant and generate electricity. Later on, in 1954, when Lewis Strauss, then chairman of the U.S. Atomic Energy Commission spoke of electricity as "too cheap to meter," I, as well as the public in general, believed that energy from nuclear fission would provide an inexpensive power source for all mankind in the future. But my encounters some years later with the

File:F-1 rocket engine at United States Space and Rocket Center in 2006

From Wikipedia, the free encyclopedia

bureaucracy of the Atomic Energy Administration and its champion of nuclear energy, Edward Teller, made me take a second look at the prospect for cheap and peaceful applications of nuclear power.

My first assignment at JPL was to help build a pump to move the rocket fuel from the tank to the combustion chamber. I had taken a course in pumps from Professor Richard Folsom at Berkeley, and it came in handy for this assignment. However, this job did not offer me an opportunity to do experiments, and when I heard one day about the plans to build a heat transfer laboratory at JPL, I immediately volunteered. Today, it is hard to imagine that a young engineer, just a year out of school, would be offered the opportunity to build from scratch and guide the heat transfer research for the Jet Propulsion Laboratory. The big challenge of the day was how to regeneratively cool liquid propellant rocket motors, similar to the German V-2. This required using the fuel from the storage tank, prior to its injection into the combustion chamber, as a coolant to keep the throat of the rocket motor from overheating or "burning out." We knew that the Germans had successfully cooled the rocket motor of their V2 in this fashion, but there was virtually no experience or data in the United States on how to accomplish this task. Here I was, a young refugee from Nazi persecution, just a year out of school, asked to find a solution to a problem that turned out to be crucial in future efforts to put a man on the moon. It turned out later on that the same challenge occurred in the development of so called "boiling nuclear power plants" and that my work would become a crucial decision point for the nuclear industry. At the time, I had very little appreciation, however, of the significance of my assignment, and set to work building a laboratory that could simulate the high heat flux conditions at the throat of a rocket motor. The photo shows an F-1 rocket engine similar to that used in the space program now on exhibit at the U.S. Space and Rocket Center. The tubes wound around the outer surface are used to carry the liquid coolant from storage to the combustion chamber of the rocket motor.

My professional life was going well, but I was lonely for a female companion. I was quite shy, unsure of many aspects of my personal and social life, and as a result lonely despite many professional and superficial social interactions. One day my sister Susi told me that she had run into a very interesting woman that she would like me to meet. Her name was Helen W., and at the time she was a secretary in a law office in Beverly Hills. Susi managed to introduce us and I was quite taken by Helen's appearance and manners. Helen was not beautiful, but she had a regal appearance, walked erect with a confident step, and had a wide smile showing her brilliant white teeth. She had long red hair and a light complexion with lots of tiny freckles on her nose. Helen was one or two years older than I and exuded a sense of quiet ability to manage life. It was obvious to me that although Helen had lived all of her life in the United States and had never faced the trials and tribulations of a refugee, she had undergone other deep life experiences that made her in some ways much more mature than I.

Helen wanted to become an opera singer and she had made progress in her efforts to become a professional singer. At the age of 21, she met and married a physician who also had an interest in singing. She was never willing to talk much about

her husband or any details of her relationship with him, but after they married, the doctor bought a house. From what I could learn, they had some good times together. While she was married to him, a great professional opportunity came into Helen's life. Lilly Pons, one of the great Metropolitan stars, became sick and at the last minute Helen was asked to take her place and sing Musetta in Puccini's famous opera La Boheme. She had good reviews and hoped to build a career on her success, but soon after, tragedy struck in her life. Her husband apparently had some mental problems and one night rammed his car at 100 mph into a solid stone wall. Helen told me that she was convinced he committed suicide, but since there was no suicide note, the cause of

death was officially called an accident. She was a widow at the age of 23, quite naïve about what to do in her new role, and some unscrupulous realtor cheated her out of the house that should have gone to her as an inheritance. There was little money left and Helen had to go to work in order to make a living. This ended her dream of becoming an opera singer.

Helen had a scar on her lower abdomen. When I asked her about it, she told me that sometime after her husband's death she was diagnosed with a cyst on her ovary and a surgeon removed both of her ovaries. He told her that this was the simplest way to take care of the problem. My very good friend, Paul Toch, who in the meantime had also become a surgeon, later told me that from all he knew about Helen's condition, one could have dealt with the problem by removing the diseased part of the ovaries without making the patient infertile for life. But we all know how difficult it is to oppose the opinion of an expert, especially in the medical field. At the time I only felt deep compassion for Helen, but did not give much thought to how her condition would later also impact me and our future.

After going out with each other for several months, I found that in addition to my toothbrush, most of my clothes and professional material were in Helen's apartment. In those days, people still raised their eyebrows and made comments behind a couple's back when they found out that a man and a woman lived together without the blessing of matrimony. Today we so easily accept it when an invitation proffered to a man or woman requests that they bring his or her "significant other." But at the time Helen suggested that we live together, this was an *avant garde* idea for its day. All the same, it made economic sense and eliminated a lot of wasted time driving from our family's home to Helen's place. Our friends were sufficiently broadminded to accept our arrangements without question, since they all knew us as a couple. We might have continued living together without facing some fundamental questions about our future if it had not been for my professional aspirations.

By the summer of 1949 I had been at JPL for four years and my work was going well. The challenge of building the JPL heat transfer laboratory had been successfully completed, and I was conducting interesting research in boiling heat transfer. I had published three articles on boiling heat transfer, the first of which was probably the most original and innovative of all my research papers. It was based upon research work I had done in the newly built laboratory and the interpretation of the data was subsequently verified by similar experiments conducted by Professor Warren Rohsenow at MIT. I had good technicians working for 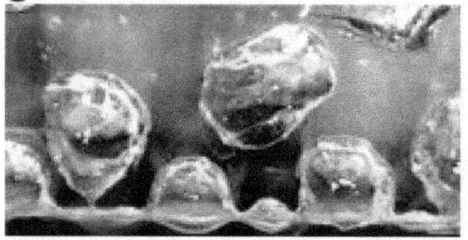 me and some military graduate students from the Cal Tech Officer Training program also participated in my research. However, there were two elements missing in my life. The first was a belief that the lack of mathematical education I had received was handicapping my research. I was aware of the complexity of the boiling mechanism in which bubbles formed at a surface heated above the boiling point and then moved into the fluid, thereby increasing the level of turbulence and also the rate of heat transfer way above what is possible with ordinary convection. As a part of the research, conducted jointly with Mr. Gunther, an officer in ROTC at Cal Tech, we had photographed bubbles and knew a good deal about their growth and decay history. I thought at that time that if my mathematical capabilities were better, I should be able to describe the life history of these bubbles mathematically and then be able to predict their behavior under various real-world conditions with any type of fluid. In retrospect, this was a naïve expectation, because boiling heat transfer is still a topic of intense research and many people are aiming for the same goal I had—an analytic description of the growth and decay of bubbles.

I had taken night courses during the time I worked at JPL and in 1949 completed my Master's degree at UCLA under L.M.K. Boelter who had left Berkeley and become Dean of Engineering for the LA campus. The topic of my thesis was heat transfer with change of phase, a promising new way of storing thermal energy. Boelter was an early believer in the future of computers and encouraged me to try my hand at this new technology. UCLA had some analog computers that required feeding them hundreds of computer cards to run even a simple program, a task with which my sister Susi helped. The thesis problem was solved by means of this crude computer and the results were published (Solidification and Melting of Materials Initially at the Fusion Temperature [with F. Romie], *Proc. Phys. Soc.*, series B,v.158:277–291,1951). But I had not had the leisure of devoting time to improving the mathematical background which I was lacking. I was, therefore, playing with the idea of going back to school for a PhD and then returning to JPL.

Early in 1949, I learned about the availability of a Guggenheim Fellowship in Jet Propulsion and I decided to apply for it in order to learn more mathematics and get a PhD. The Guggenheim Foundation had established two centers for jet propulsion, one at Cal Tech and one at Princeton. A few weeks after I had put in my application, I received the long-awaited letter from the Guggenheim Foundation. I was elated to learn that I had been granted one of the four fellowships and had my choice either to go to Princeton or to Cal Tech. I decided to go to Princeton because, given all of the circumstances of my life, I felt it would be a good idea to change my venue and try something entirely new. I was not sure whether or not Helen would come with me, and my friends later told me that they thought I really wanted to go by myself. But Helen assumed that our relationship was so close that she took it for granted that we would go together. There were times in my life when not to act seemed easier than searching out my true feelings to find the right course of action. This was one of them, and, willy-nilly, Helen and I made plans for our trip to the East.

I owned at the time an old jalopy that I did not trust to make the 3,000 mile

trip. Since I had been working at this point for several years at a relatively well-paying job, I had saved enough money to buy another car. Given my elation about the forthcoming trip and needing good transportation, I decided to buy a new car, a light blue Chevrolet two-door that seemed, at the time, like a dream come true. Helen and I took off at the beginning of August to travel to Princeton. Helen's father had worked for Conoco, and for a time he had been stationed in Colorado. Helen therefore knew Colorado and when we arrived at the Rocky Mountains, she showed me a new world. Ever since I left Austria, I had in my memory the wonderful Austrian Alps, and here before me spread a glorious chain of mountains that in its monumental grandeur rivaled the Alps. We camped and hiked, and the most memorable of those hikes was a climb up Long's Peak. We took two days for the climb, spending a night halfway up at Chasm Lake. The second day, the weather was good and we arrived at the peak sometime before noon. At the time the climb was easy because permanent ropes were installed for the most difficult part just before reaching the top. Long's Peak is the highest of the mountains in the vicinity and the view from the top is unforgettable.

Our trip was time out of time. There seemed to be nothing to interfere with our interaction. I am not sure it was exactly love, but it was certainly a closeness and a sense of well-being from sharing experiences such as hiking up a mountain, swimming in a lake, watching sunsets, and sleeping together in the open or in a tent that we had brought along. A week or ten days later, we finally arrived in New Jersey and crossed via the Manhattan Bridge into Manhattan. I stopped the car in the middle of the bridge and could not help but shout out aloud something like, "This is incredible." And indeed, the feeling was one of incredible joy. Just about nine years before I had arrived here as a penniless refugee and now I was crossing into New York in a new car with a beautiful woman and expecting to finish an education which had been interrupted several times. Little did I know that having this beautiful car would eventually be my undoing. But, for the moment, there was nothing to spoil my elation and joy and I will forever remember that incredible feeling of having fulfilled the American Dream. It was a day or two later when I arrived at Princeton that reality took over again.

CHAPTER 11—PRINCETON UNIVERSITY, 1949–1951

Soon after our arrival in Princeton in September of 1949, Helen proceeded to New York, where she found a job as a secretary in a law firm. She shared a small apartment with another woman and it was not possible for us to meet quietly in New York. Since I had a private room in Princeton, she came to visit me. When the landlady saw her leaving one morning, she asked who it was that she had seen. I told the landlady that we were married. I am not sure the landlady cared, but this white lie was to haunt me later on in unexpected ways. Our meetings continued for the next two or three months, when I realized that I could not imagine a permanent union that would not give me children of my own. We tearfully said goodbye to each other and, soon

thereafter, I moved into the student dorms on campus.

After I came to Princeton in September of 1949, I moved into a room on top of the garage in the home of an English professor on Prospect Avenue. With me in the house were two other Guggenheim Fellows. One was a Chinese man by the name of Chang and the other was an army major who had come from Harvard by the name of Frank Bailey. Both of these men became quite friendly with me. Chang obtained his PhD at Princeton some years later and then was named to the faculty, where he achieved the rank of full professor. I lost track of Frank Bailey, unfortunately. There were altogether seven students working towards their PhD in the department. In addition to Dr. Chang, one of the other students, Dr. Ronald Probstein, got his PhD and subsequently became a full professor at MIT. After a few months, a vacancy occurred in the graduate dormitory and I moved onto the Princeton campus.

My status at Princeton was rather strange because I had essentially regressed from being an instructor at a university and a project engineer in charge of a major laboratory to that of a student. The adjustment period was very difficult for me, partly because I had apparently alienated one of the instructors. From that time on, the instructor was quite antagonistic towards me and I could never understand why. It was not until many years later that I learned the reason for this unfortunate situation. Apparently, when I first drove into Princeton, the wife of the instructor saw me in a new car and complained to her husband that here was a Guggenheim Fellow able to afford a new car, while their family had a much older vehicle. Another reason for my problems was that the majority of the instructors in the aeronautics department at Princeton were on temporary appointments because they did not have PhDs. In fact, only one of the instructors, Dr. Joseph Charyk, had earned a PhD in Canada. All of the others had master's degrees, just like I did, and were teaching mostly due to the fact that they had a position in the research establishment of the university and were available on temporary assignments until qualified faculty could be recruited. The man in charge of the aeronautics curriculum was an Italian engineer by the name of Luigi Crocco, whose father had been a general in the Italian Army and had pioneered some experiments in rocketry. Luigi had gone through an Italian university and had also participated in some early rocket research. As far as I know, he also did not have a doctorate. Consequently, it was a strange situation where a group of instructors with master's degrees and less experience than I had were in charge of a program consisting of six or seven young students working towards their PhD. This was, of course, only a temporary arrangement. During my second year at Princeton, an intense recruitment effort for qualified faculty was initiated and two years later the university had an outstanding faculty in the aeronautics department. But by that time I had already left.

At the end of the first year, I learned that I had successfully passed all my courses. Although my grades were not brilliant, they were sufficient to be given a second year of the Guggenheim Fellowship. The courses for the second year were actually not as difficult as the first, except for one highly mathematical course taught by Dr. Charyk. All the students agreed that passing the qualifying examination necessary to enter the final phases of the PhD program, it would probably be the questions in the mathematical area that would determine the pass or fail. The second year at Princeton was quite uneventful as far as my studies were concerned, but I realized that I did not really have an aptitude for highly complex mathematical analysis. My abilities and interests were more on the practical side—doing assessments of technical problems and working in the laboratory doing experiments that required imagination and perseverance. I did pass all of the courses at the end of the second year, but an event prior to the end of the semester really was far more eventful than my studies. That event had to do with my security clearance to engage in classified work for the U.S. government. Since this event shook up my entire life from here on, I will describe it in detail in the next chapter.

In May of 1950, I learned that my father had suffered a heart attack, a coronary thrombosis, while working as a gardener. He managed to walk back home, but was immediately taken to a hospital where he was treated successfully. This event obviously upset me greatly, because now my family had lost the main breadwinner. Although my father recovered from the heart attack, he became unable to continue working as a gardener. I could hardly wait to get through the final exams and rushed home by automobile as soon as possible thereafter. One of my Princeton friends, Ken Scott, accompanied me on this trip. We took turns driving and managed to cover the entire 3,000 miles in about 40 hours. When I arrived home, my father was able to get around and, with the help of a medication called Rauwolfia, was able to return to a more or less normal life. But the doctor prohibited him from any heavy manual labor, such as gardening.

CHAPTER 12—DENIAL OF SECURITY CLEARANCE, 1950–1953

"A thing long expected takes the form of the unexpected when at last it comes" –
Mark Twain

I have not thought about this chapter in my life for many years. However, the CU
retired faculty book club to which I belong at one time selected a book entitled
American Prometheus—The Triumph and Tragedy of J. Robert Oppenheimer by Kai Bird
and Martin J. Sherwin. Reading this book brought back to mind events in my life that
almost paralleled those of J. Robert Oppenheimer, whom I had previously met in 1945
at Cal Tech, soon after I graduated and began working at the Jet Propulsion Laboratory.
Ever since that time, I retained the hope that nuclear fission could one day be used for
the betterment of mankind and hoped to be able to work towards that goal. I was
therefore pleased when, in March of 1950, I received a call from John Menke, who
identified himself as the President of the Nuclear Development Associates (NDA) in
White Plains, New York. The staff of this organization acted as consultants to some of
the large industrial corporations such as Babcock and Wilcox, Cooper Bessemer
Corporation, and Industrial Nickel Company on heat transfer problems. He asked if I
would be willing to become a consultant with NDA to help with some energy-related
work and then accept full-time employment for the summer of 1950. I accepted this
offer and, after my visit with my father, I returned to New York where I worked at NDA
until school started again in September of 1950. When I came to NDA, Mr. Menke told
me that his organization had received a large contract from the Atomic Energy
Commission to investigate the safety and feasibility of a new type of reactor for a
nuclear power-plant and NDA was interested in using my JPL experience in boiling heat
transfer. The goal of that project was to investigate the feasibility of operating a
nuclear power plant under conditions which are referred to as surface boiling, a
condition of boiling that I had investigated at JPL. That means that the fuel rods, which
consist of uranium and generate the heat for making steam and running the power
plant, were not cooled by ordinary forced convection. The heat generation in the
nuclear rods was permitted to increase to a point where the surface temperature
exceeded the boiling point, while the bulk of the fluid remained below the boiling
point—that is, a condition called sub-cooled boiling. In order for this scheme to
operate safely, it was necessary to know the volume occupied by the vapor in the

bubbles generated by the boiling condition in the boundary layer near the surface. This was a problem that I had studied at JPL in connection with the cooling of rocket motors where similar conditions prevail. Hence, I was one of the few people in the country with the experience in the key issue of this design and also considered this to be a great opportunity to apply my engineering knowledge to the peaceful applications of nuclear energy. I was thrilled to be able to apply my knowledge and experience to such a significant challenge.

As a part of employment with NDA, I was asked to apply for Q clearance from the Atomic Energy Commission because some of the information pertinent to NDA's work was classified. I was told that, in accordance with the Atomic Energy Act, an investigation would be conducted concerning my character, associations, and loyalty. At the time, I had absolutely no qualms about subjecting myself to such an investigation because not only did I consider myself a loyal citizen of the United States, but I had also been investigated on two previous occasions: once in 1943 for my machine shop work in Los Angeles, and again in 1945 in connection with my research at the Jet Propulsion Laboratory. During the second investigation for Army/Navy clearance, ten or eleven people were visited by investigators and, in a summary letter, it was stated that no derogatory

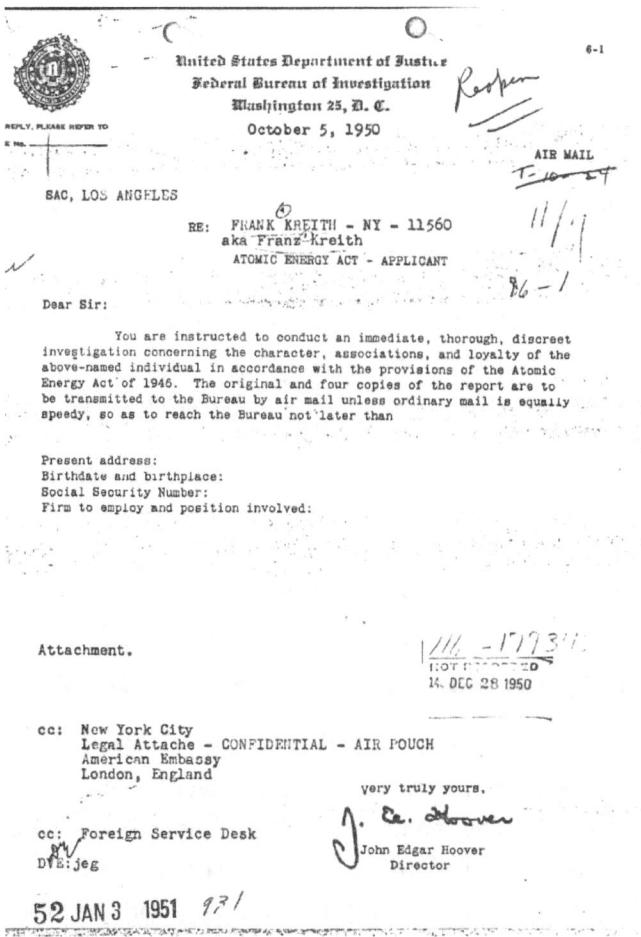

information was obtained. Other people were contacted by letter, including Mr. Schigut, who had been my first employer after I arrived in California in 1941. He wrote a letter of recommendation about me to the Army in answer to their Inquiry No. LA18005, dated July 1945. Following this investigation, I received Army/Navy clearance to work at the Jet Propulsion Laboratory up to a level of Secret in the latter part of 1945. But the letter from FBI Director J. Edgar Hoover in January of 1951 contained a warning that was not what I naively expected to hear.

Having twice before been cleared of any disloyalty, I was naturally stunned when in the middle of the school year I received, on January 31, 1951, a letter from the AEC informing me that

"The substance of the information provided by the investigation which raises a question concerning your eligibility for employment on Atomic Energy Commission work, is as follows:

"It was reported in 1941 that you, while employed in Culver City, California, passed out pamphlets advocating a rule by the masses and the Communistic form of Governmental control. It was further reported that you read various literary works of Marx and Engels and principles advocated by these writers

It was reported that in 1942, while you were employed in Hollywood, California, you made statement to fellow workers that they should not work, inasmuch as they were not being paid as much as the owner of the company; that they should not work because Russia was going to take over America anyway.

It was reported that in 1944, you attended a party sponsored by the American Youth for Democracy. The American Youth for Democracy is listed by the Attorney General as an organization coming within the purview of Executive Order 9835.

It was reported that during the period from 1947 to 1949, you associated with Dr. David Altman who in turn has been reported to be very friendly with Sydney Weinbaum. Sydney Weinbaum, a reported former member of the Communist Party, was convicted on charges of perjury and fraud against the government and was sentenced on September 12, 1950 by the U.S. District Court, Los Angeles, California."

I found the accusations contained in the letter ludicrous and answered them in a letter dated February 6, 1951. In that letter, I denied absolutely ever having passed out pamphlets in Culver City advocating rules by the masses and a Communist form of government, or having made statement to fellow workers that they should not work because they were not being paid as much of the owner of the company and because Russia was going to take over America. I did admit to having read parts of the works of Karl Marx, but pointed out that I had also read standard texts by Adam Smith, John Stewart Mill, and others. I also admitted to having attended a party sponsored by the American Youth for Democracy, but also pointed out that I had no political affiliation with that group and that it had not been listed as among subversive organizations until after 1944. I readily admitted to playing bridge and having social interactions with Dr. David Altman, but had never met Mr. Sydney Weinbaum in person.

At the same time, I also asked certain questions to help me understand in more detail where the accusations originated, particularly where and to whom I was supposed to have passed out the subversive literature and discouraged workers from working, when and where I was supposed to have attended a party sponsored by AYD, and what specific suspicions were raised by my social association with Dr. David Altman. On February 13th, Mr. W.E. Kelly, the Manager of Operations of the AEC, wrote me that he "regrets to be unable to furnish any particulars about the derogatory information that I sought because the commission was anxious to protect its witnesses and avoid compromising sources of confidential information." In a letter dated February 15th, which I received on the afternoon of February 20th, Mr. Kelly informed me that a personal security board had been appointed and that my case would be heard in New York City two days after I received the letter. The board was to consist of a lawyer, a professor of chemistry at NYU, and a member of the U.S. AEC. In my letter of February 21st, I pointed out that I had been seeking information that could not be obtained rapidly and the AEC agreed to delay the hearing until April 13, 1951.

When I came to the meeting, I was aware that the members of the board had been given all sorts of information from the FBI to which I had no access. At the same time, I had heard about investigations by Senator Joseph McCarthy of former Communists and Communist sympathizers in which the ticket to freedom was to be a good stool pigeon and inform on other people. When I entered the meeting, I had made up my mind that under no circumstances would I divulge any names, because it was apparent that if I mentioned a name or pointed a finger, it would cast suspicions that could lead to serious problems for that individual. I realized that this position might be detrimental to me, but I felt so confident of my innocence, of not being a Communist, and the falsity of the accusations, that I was willing to take this risk. The people who had brought the charges against me may not have been aware of the implications of the statements they made to the investigators and I did not want to have it on my conscience to have created similar problems for others in an effort to placate the members of the investigative board. In respect to my previous girlfriend, Doris, who had told me that she had been a member of the Young Communist League, I felt that the FBI had a full record of who belonged and who didn't, and really did not need me to tell them. However, I also suspected that volunteering this information would have been a test of my "loyalty." In any case, during questioning by the board, I neither named any names nor engaged in any arguments with the investigators, but simply tried to stick to the facts relating to the derogatory information. I am today still proud of my decision regarding this matter.

On July 31, 1952, I received another letter from Mr. Kelly informing me that security clearance had been denied. The rationale was that during the hearing I did "not show the candor and truthfulness of one seeking to resolve questions on security clearance," that my "veracity and integrity are questionable," and that I "appeared to be a person of poor moral character." The letter further stated that my claim to have left Mr. Schigut's employ in order to get a better job was in contradiction to Mr. Schigut's claim that he fired me, and therefore my entire testimony was "open to serious doubt." During the hearing, I had admitted to passing out pamphlets on anti-Semitism and labor union activities, and the Commission concluded that there was no evidence that any of the literature I gave to fellow employees had anything to do with communism. Nevertheless, the Commission did not absolve me of having advocated Communist ideas. On another point, the FBI had discovered that in one of the employment questionnaires I had previously filled out, I did not mention employment (of less than a month's duration) at a small machine company called Leckinger. This omission was interpreted as an effort to deceive the investigators. Finally, I had not

been able to recall two names of fellow students during the interrogation. The interrogators had cited these two names in passing, and asked whether I knew these people, without telling me when, where, or under what

SAC, New York — October 10, 1951

Director, FBI

FRANK KREITH, aka
Franz Kreith
HELEN WEATHERFORD - VICTIM
WSTA

Reurlet 9-26-51 entitled Frank Kreith, aka, AEAA.

The facts regarding the possible White Slave Traffic Act violation on the part of the subject should be discussed with the USA, SDNY to determine if prosecution would be considered. It should be pointed out that the violation appears to be in the nature of a personal escapade.

EX - 6

RECORDED

EHH:lmo

circumstances I was supposed to have known them. Later it turned out that one of the names was cited incorrectly. The board also questioned whether I had testified "fully and frankly" regarding my association with AYD, although I had admitted attending some meetings sponsored by that organization. They further claimed that because I had listed myself as a consultant to Princeton University, this statement was an attempt to deceive. Princeton later confirmed that I had been a consultant to Project Squid, which operated under a Princeton University charter.

The final accusation was one to which I frankly admitted. It pertained to my use of the word "wife," when questioned by my first landlady in Princeton, as it applied to my relationship with Helen. As Helen and I had lived together in Los Angeles in a "common law marriage," I did not feel that this response to my landlady's query was dishonest. At the hearing I admitted of course that I was not legally married to the woman and the Commission therefore concluded that I "was a person of poor moral character" and could be subject to having violated the White Slavery Act, which prohibited bringing a woman across state lines for sexual purposes. The letter of Oct.10, 1951 to the FBI Director J. Edgar Hoover asking for my possible prosecution is shown above. Today this would be a laughable accusation, but it seems that the FBI was seeking any means of tying me to some illegal act, no matter how farfetched.

Although the Commission agreed that there was no evidence that I had ever advocated communism, passed out Communist literature, made statements advising fellows that they should not work, or had any associations with Sydney Weinbaum, the members nevertheless decided to deny me security clearance. At this point, I realized that the outcome of this security investigation could have a permanent impact on my life and I made another serious attempt to try and set the record straight. From a question at the hearing, I suspected that the main accuser was Mr. Schigut, and I requested that the Commission obtain a copy of the letter of recommendation that he had sent on July 9, 1945 to the investigators that approached him in connection with my Army/Navy clearance, which had been subsequently granted. Mr. Schigut promised to check his files, but nobody was surprised when he said that he could not find the letter. I then wrote to the Armed Forces Ninth Service Command in Washington, DC, on August 4, 1952, requesting a copy of the letter that Mr. Schigut had written about my loyalty, but received no reply. It was, of course, odd that Mr. Schigut had no derogatory information about me in 1945, but seven years later stated with certainty that he knew I was a Communist.

After having refuted in writing all of the specific charges regarding my loyalty, my lawyer thought there might be another chance to clear my name. He suggested that there be a direct confrontation between me and Mr. Schigut in Los Angeles. The Atomic Energy Commission agreed to have a third interview, which took place on June 6, 1952 at the Atomic Energy Commission office on Wilshire Boulevard in Los Angeles. When I came to the interview, Mr. Schigut looked at me and said, "You look nervous." I responded to him and said, "I would not have to be nervous if you would only tell the truth." He answered that "I can tell a Commie when I see one and you are one of them." When we entered the room and Mr. Schigut began to testify, it turned out that he had become even more aggressive. He now claimed that Frank Kreith was an admitted Communist, that he had seen Frank Kreith's Communist Party card, that Frank Kreith distributed Communist literature in the plant, and that Frank Kreith openly advocated the principles of the Communist Party. At the beginning of the hearing, Mr. Schigut was asked whether he would be willing to testify under oath. He said at first that he would not take an oath because he was afraid that I might file a charge of slander after the meeting if he repeated his adverse opinions. But once he was told that whether or not he was under oath would make no difference, that a charge of character assassination could only be filed depending upon the statement that he made, he was willing to take the oath. However, under questioning by the counsel, he admitted that his assertion that I was an admitted Communist was simply based upon his opinion. I pointed out to him that I had shown my Machinists' Union card from England to him when I started employment and wondered if he could have mistaken that for a Communist Party card. He said he could not remember details. When asked whether he actually saw me distribute Communist literature, he said no, he did not, but another worker brought a leaflet to his office and he remembered that the headline was "Workers Arise." He therefore assumed that it came from me. When asked further as to the details regarding my advocating principles of the Communist Party, he admitted that his hearing was bad and he did not wear a hearing aid at the shop and, therefore, his recollection was very vague. He still maintained that he had discharged me because of ineptitude, but also admitted that some of the stamping machines were lacking the safety guards and that this had prompted me to start looking for another job—particularly since the two missing fingers on Mr. Schigut's hand were the result of an accident with a heavy stamping machine of the kind that I was asked to work on. I also testified without rebuttal from Mr. Schigut that when I refused to work on that machine without the guard, I was transferred to some other job. The only member of the Commission present at that hearing was Mr. Lawrence Berman. He was a lawyer and, at the end of the hearing, I was cautiously optimistic that Mr. Schigut's testimony had been properly rebuked. My life became very full after the hearing and I could do nothing but wait and hope.

CHAPTER 13—MARRIAGE TO MARION, AUGUST 22, 1951

I left Princeton University at the end of May 1951, just as soon as my final exams were over. When I left Princeton, I think I was vaguely aware that I might not return to complete my PhD. After all, I had an offer to teach at the University of California starting in the fall and the offer mentioned nothing about my having to complete the Princeton degree. Professor Crocco had mentioned that he wanted me to do an experimental thesis and there was really no equipment at Princeton with which one could begin to start an experiment. Hence, his idea was rather vague and probably based mainly on the fact that I did not have the qualification or interest to do a theoretical thesis. But the real obstacle was the comprehensive examination for which the rest of the students began to study just as soon as the semester finished. Having the offer of a job to teach at Berkeley gave me a different perspective and I felt a great sense of relief to be able to leave Princeton and go back to California.

In departing from Princeton I was also leaving behind Doris, a woman with whom I had had an intimate relationship for many months. She was a nurse living in New York and we met almost every weekend, either at her apartment or at my room in the graduate college. Although Doris was good looking and we were physically very attracted to each other, I did not think that our relationship would ever lead to marriage; whereas, Doris assumed that it would be permanent. My reasons for the hesitancy in forming a more permanent relationship were that our backgrounds could not have been much more different. She was brought up a Catholic in a very strict home, had an upbringing that one would almost consider WASP-like, and never had experienced the kind of trauma that comes with being a refugee. We had good times together, but never really formed the kind of deep understanding that I thought necessary for becoming permanent life partners.

I took off from Princeton in my convertible Chevrolet and arrived in Los Angeles in the beginning of June. There were a lot of people in LA anxious to see me and I soon settled into a new life free from the anxieties of studies at Princeton. I had pretty much decided not to return in the fall to take the comprehensive examination, and I began to enjoy the freedom of California. There was of course still a third interview with the AEC Commission looming in the foreground, but I was at that time still optimistic about its outcome. My father helped me prepare for the meeting and the clarity of his thinking gave me an inkling of what an outstanding lawyer he must have been in Vienna. There were, of course, also many old friends as well as the promise of seeing Susi and Kurt that buoyed my spirits. My parents did not have many social contacts, but they mentioned that they had met another family by the name of Finkels who were also refugees and lived not far away on Havenhurst Avenue. My parents were still living in the house at 722 Westborne that we had bought many years before when I was still at JPL. But, in addition to both families being refugees from Hitler and living in close proximity, there was another factor in my parents' interest to have me meet the Finkels. The Finkels family had two daughters, Marion, who was 24 at the time, and a younger sister, Ada. I know that my parents had met the daughters on several occasions and were, of course, anxious to have me meet them as well. I do not remember when and where I first met Marion, but it was undoubtedly in connection with our families' interaction. I do remember being introduced to Marion sometime in June. She was working at the time as a secretary to two psychiatrists in Beverly Hills. One of them was an older man. The other was in his late 30s and he owned a small yacht. Going sailing on a private yacht seemed like a wonderful experience and Marion was able to arrange to have her boss invite both of us for a trip out of the Los Angeles harbor. It was the first time that I had been sailing on a private yacht and I appreciated Marion's ability to wrangle this invitation.

Soon after we met, Marion also introduced me to some in her circle of friends. They were a vastly different type of people than I had been associated with in my engineering career.

Most of them had a background in arts and sciences from the university, were interested in literature, and many of them were aspiring to become writers for the movie industry. They knew some of the movie writers and directors who had escaped from Germany and were slowly beginning to make a name for themselves. Marion also had an on-and-off boyfriend by the name of Shimon who wanted to write for a serial in Hollywood. It may have been a precursor of the LA Law series. But, the most remarkable person that Marion introduced me to was her cousin, Baruch. He had immigrated to Israel and was beginning to participate in the fledgling movie industry of the country. I believe that he also knew a number of people in Hollywood and was a very good friend of the famous movie actress Shelly Winters. Seeing and talking to Baruch was also an exciting experience because he had the ability of describing what he was doing so that you would actually become a part of his activities. At one time, he asked us to come up with a title for a movie he was producing and I recall that we were not very successful. Knowing these friends of Marion's opened a door for me because they were looking at parts of the world that were new to me. Marion's parents came from Altenau, a small town near Hamburg. Her father owned a large department store, while her mother came from a poor Polish family. She was a very beautiful and capable woman, whereas her father was a small reticent man with few virtues other than being kind and considerate. I later learned that Marion's mother had essentially taken over responsibility for running the department store. Marion had photos of their apartment in Altenau and they show a taste for modern and expensive furniture. When Hitler came, the family lost the department store and, with the help of Marion's Uncle Joe, managed to immigrate to Belgium. Joe had friends all over the world, was able to salvage much of his wealth, and started a successful garment factory in Canada. He must have, at one time, had a bad accident, because one of his legs was shorter than the other and he could not bend it. Despite this physical handicap, he was an impressive man and I came to admire him greatly. The story of Marion's family's escape from Europe would make a wonderful book unto itself. Marion has written up parts of the harrowing escape and I include only the major details here. In summary, the family escaped to Brussels two or three years after the Nazis took over Germany. Marion attended high school there and managed to become fluent in French, a language that she has loved ever since. I do not know much about their stay in Brussels, but believe that it was fairly pleasant and relatively uneventful. All of this, however, came to a crashing halt when the German Army invaded France with a pincer movement, passing through Belgium. The Finkels family and an aunt and another person managed to get a hold of a small automobile and were able to arrive in unoccupied France, barely escaping the German Army. After arriving in France, Marion's father was sent to an internment camp and the family began frantic efforts to escape from Europe. The only country for which they could obtain a visa, probably with the assistance of Uncle Joe, was Cuba, which at the time was under control of a dictator by the name of Batista. The family expected to stay in Cuba for a short while and then go on to the United States. However, their stay lasted for almost five years,

and Marion became the main breadwinner for the family. A number of wealthy industrialists in the diamond industry were allowed to come to Cuba, provided they would bring with them the knowhow to make machinery necessary for processing raw diamonds into valuable stones. For this work, they required nimble hands and these were provided by youngsters such as Marion. Neither of her parents could find any work in Cuba and, therefore, Marion was forced to work for the entire four-and-a-half years that the family spent in Cuba. She tells me that she learned how to girdle small diamonds, which were then used to make jewelry.

During that time, Marion learned to speak and communicate in Spanish, but was not able to continue a formal education. The family was finally able to come to the United States in 1946. After they arrived on the

East Coast, there was a race between Uncle Joe in Montreal and a cousin in Los Angeles as to who would first find an apartment for the Finkels family. Los Angeles won out, and the family moved to LA in 1948, I believe. Marion had learned to type and took a secretarial position with an organization called the Council of Jewish Women. This organization attempted to help Jews escape from Europe and also reunite families that had escaped but had lost touch with each other.

There was obviously a lot of similarity in Marion's and my backgrounds. We were both Jewish refugees, but neither of us was very religious. We were proud to be Jews and did observe some aspects of the High Holidays, but did not make that a fetish. We both came from well-to-do parents, although I think that Marion's parents were much wealthier than mine. The saga of immigration had interrupted the schooling for both of us, and although Marion attended some evening classes at the Los Angeles City College, I do not believe that she ever attained a formal high school diploma in the United States. Both of us had very powerful mothers, although Marion's mother retained her appearance and influence, while my mother badly deteriorated in stature after coming to the United States. I think Marion's mother admired my father and I know that my father thought highly of her. I must have stood very high in Marion's mother's eyes, because to be teaching at a university was highly esteemed in Europe. I would not say, by any means, that ours was an arranged marriage, but

certainly our meeting was arranged by our parents. After meeting each other, Marion and I had a number of dates, did a lot of fun things together, but never seriously thought about the future. However, the future caught up with us, because I had to go up to Berkeley to find a place to live and make arrangements to start my teaching career. At that point, we knew a lot about our respective pasts, but I do not think we really had a deep knowledge of each other. At the same time, we were at the point in our lives where to be married was a natural step to take. Today nobody would bat an eyelash if we had decided to move in with each other and begin to learn more about ourselves, but this was not an acceptable option in those days. So, here we were, attracted to each other, aware of many similarities in our lives and value systems, with parents most anxious to see their daughter and son be married, but forced to leave each other, at least temporarily because Marion had a job in LA and I had to go to Berkeley. I am not sure whether or not I asked Marion to come to Berkeley to live with me, but if I did, I am sure that her reply would have been, "Not unless we are married." After 56 years and retrospection, I would say that our decision at the time to get married was not based on a rational and reasonable foundation. However, that was the decision we made and being practical people we decided to meet halfway between LA and Berkeley to take this final step. Marion had a couple of friends who also, at the time, lived apart—Lottie in Berkeley and Allan in LA. Once again, practicality took over and I took Lottie in my car, while Marion took Allan in her car, to Monterey where the four of us met. Marion had $35 and an old Studebaker and I had about $250 in a bank account plus a convertible Chevrolet. That is how we started our life together.

CHAPTER 14—UNIVERSITY OF CALIFORNIA, BERKELEY
1951–1953

After arriving in Berkeley, Marion and I found a small apartment high up in the local mountains. From one of the windows, you could actually see the lights of San Francisco and the Bay Bridge connecting San Francisco to Oakland. It was a busy time for two newlyweds because I had to prepare for teaching university-level courses for the first time in my life and Marion was able to fulfill her dream of going to college where she enrolled in the French department. We were the youngest faculty couple and all of my colleagues at Berkeley were much more experienced with university life and academic politics than I was. At the same time I was hired, the university also brought in two other professors. One was Ernie Sparkman who had graduated with a B.S. and M.S. from Berkeley and was working as a research engineer at the Shell Oil Company in San Francisco. Ernie had quite a few friends on the faculty and was able to connect with them easily. He was a short, slightly built man with a crew cut and always dressed impeccably with a bow tie. Some years later, he was elected chair of the department, but then left the university to become vice president for environmental affairs at General Motors. The other person hired at the same time as I was a chemical engineer by the name of Paul Stewart. Paul had earned a PhD in chemical engineering at the University of Texas. He was a very kind man, leaning towards the obese, with a face that easily smiled under his round spectacles. He remained at the university for many years and retired in 1975.

My first year in academia was fairly uneventful. I was asked to teach a course in thermodynamics and since I was the lowest on the faculty totem pole, I was also given responsibility for teaching two identical junior year laboratory courses. I was unaware of the fact that in addition to teaching, faculty members were expected to conduct research and obtain external funding for this work. My naïve belief that a university teacher should put his main emphasis on teaching did not meet all of the expectations of a first-rate university. However, I must have done a fairly good job of teaching because at the end of the year I was promoted to acting assistant professor. Naturally, I was elated to have the title of professor on my name. When I asked why I had the word "acting" before my name, while Ernie Sparkman was simply called an assistant professor, I was told this was a formality to be later rectified. I was unaware at the time of academic politics and simply continued to put my heart and soul into teaching undergraduate engineering students. I still have some of the detailed lecture notes I prepared in those days.

The second year at Berkeley started quite uneventfully and I continued my teaching routine. I did begin thinking about doing some research and wrote an article with Professor Steward on using a rocket motor for chemical processes. But after the first semester passed, I began to enquire about the possibility of being appointed regular assistant professor with the Chair, Professor Richard Folsom, who had originally hired me. I also began to look into opportunities for getting funds to do research in rocketry, but noticed that for most of the project, it was necessary to have security clearance. I therefore continued to hope that the security question would be clarified soon so I could also apply for funds in the classified area of federal R&D.

My battle for clearance with the AEC ended on January 21, 1953 when I received a final letter informing me that the charges against me had not been disproved and the Deputy General Manager of the Atomic Energy Commission had concurred that security clearance be denied. Hence, I could no longer participate in any work of the Nuclear Development Associates that involved matters concerning nuclear power. An extension of this decision was that no company involved in classified work could hire me. This final decision came about ten days after the University of California at Berkeley, where I had been an acting assistant professor the previous year, decided not to reappoint me for the following academic year because, as Professor Folsom stated in his dismissal letter, "My name did not appear on the roster for the next academic year." Privately he told me later on that I would not receive a tenure track appointment unless I completed my PhD degree. Needless to say, the combination of having once again lost my job and the knowledge that it would be exceedingly difficult to find another job in industry when the FBI files indicated that I was untrustworthy, would have a major impact on my future. As many refugees living under a cloud of suspicion by the administration of one's new home country, I lived from here on with a subconscious state of fear of not having a job.

CHAPTER 15—LEHIGH UNIVERSITY, 1953–1959

I am still amazed that Marion and I were not totally discouraged, but started immediately writing letters to at least 25 different universities in eight states asking for an academic position for the following year. The security investigation had depleted all of my earnings. I owed money and now was also without a job. About the same time, I received a letter from Princeton University informing me that I would have to take my written comprehensive examination if I wanted to continue to work towards my PhD. I am not sure whether or not I answered that letter. But I did know at that time that I was neither psychologically nor physically in a position to pass a rigorous three day examination. So, I continued my job search with Marion's help and finally was asked to interview at Lehigh University in Bethlehem, Pennsylvania for an assistant professorship starting the following academic year (1953/1954). I knew very little about the university or the town, but felt fortunate to have found a university willing to give me an opportunity to continue my professional life.

This was probably one of the hardest times of my entire life. I was desolate and could not see where I could go from here. In retrospect, however, it may have been a fortunate event. I was thinking how I could continue to be involved in the field of energy, but realized that the opportunity to work on nuclear power would, from now on, be impossible. I had written my master's thesis on phase change energy storage, which could be quite useful in application of solar energy, as it is an intermittent energy source. I therefore began to think about what opportunities there might be for using solar energy in the future. Although it was still sometime before I could really apply myself seriously to working on solar energy, this could have been a turning point in my life and the decisive factor that shaped my future commitment to the field of renewable energy, which at the time was only a gleam in the eyes of a few futuristic thinkers.

Lehigh University was a private school with a good reputation for its engineering education. It is located in Bethlehem, Pennsylvania, which at one time was the headquarters of the great Bethlehem Steel Company, one of America's industrial giants of the 19th and early 20th centuries. Bethlehem Steel supported and endowed Lehigh University generously. The Civil Engineering Department was the largest and best known department at the university. It was the only department that was heavily involved in research, and the field of metallurgy was a part of civil engineering, which was a rather unusual arrangement. The Civil Engineering Department operated the Fritz Laboratory which assisted Bethlehem Steel in much of their R&D. However, when I arrived in Bethlehem in 1953, international competition in the steel industry had taken a heavy toll on Bethlehem Steel and it was beginning to eliminate many of its production facilities.

Lehigh University is located on the west side of the Lehigh River on a small wooded hill. Bethlehem Steel is located just across the river and an old stone bridge connects the two banks, leading directly from the steel company to the university. Bethlehem is located in the Amish part of Pennsylvania. Just a little ways out of town one can see the Amish farming communities centered near Lancaster. Bethlehem has a long history connected to the Hussites, a religious group from Moravia (now Slovakia) who immigrated to the U.S. in the early 1800s to escape religious persecution. The only Moravian university, a small liberal arts school, is located in the center of Bethlehem across the river from Lehigh University.

I was received with open arms by the established faculty members in the Mechanical Engineering Department at Lehigh. The faculty at the university was beginning to change from mainly engineering to more liberal arts, including English, Business, and Philosophy. However, the traditional outlook of the faculty still dominated the university. In fact, the faculty was essentially divided into two parts: The older faculty members had made Lehigh their permanent home and held tenured positions. They were mostly interested in teaching and did little or no research. The younger faculty members came to Lehigh in order to get started in academia, but almost every one of them considered the move as a mere stepping stone—a halfway station on the road to a position at a more prestigious university. I was the only faculty member in the Mechanical Engineering Department who did any sponsored research and the administration rewarded me by promoting me from assistant to associate professor with tenure after my second year. Moreover, I was awarded the Lehigh Medal for Outstanding Teaching. The most important aspect of teaching at Lehigh, however, was that I was given the opportunity to develop and teach a new course in heat transfer. This also gave me the opportunity of beginning to write a textbook for a course on this topic.

Heat transfer is an integral part of most energy generation technologies. At the time there was almost universal agreement that the next energy generation technology would be nuclear power. Enormous amounts of money from the government as well as from private industry were put into developing and building nuclear power plants and several universities started a Department of Nuclear Engineering to train engineering students for a future in the nuclear energy field. Although I knew that I could not do any research in the nuclear field, I still retained the hope that nuclear power could become a significant and peaceful application of nuclear fission to atone for the destruction it had wrought as the most terrible weapon ever used by mankind. In the summer of 1958, the NSF sponsored a short course at Cornell University designed to train engineering professors to present nuclear energy courses at their home universities. I accepted an invitation to participate in the training course and learned about this new field. The course was taught mostly by nuclear physicists, many of whom had worked at Los Alamos on the development of the atomic bomb. I spent three weeks in intensive training to prepare me to teach a course in nuclear power and I presented a nuclear power course three years later at the University of Colorado. Renewable solar energy was at that time still only a glimpse in the eyes of a few futurists.

I belonged to the group of faculty members who considered Lehigh as a stepping stone. We never thought of buying a house and lived in a rented apartment not too far from the university. My teaching load was quite heavy and life was busy. Marion began to take some classes at the Moravian college and we developed a somewhat eclectic social life, split between the older faculty and some of the more exciting younger assistant professors. Our best friends were couples in the College of Art and Science, not from Engineering. They included from the English Department Professor and Mrs. Paris, an assistant professor in Philosophy of Science, Adolf Gruenbaum and his wife, and a low-level administrator, Preston Parr and his family.

At 10:41 the telephone rang
"You have a little boy" they sang
He was told that everything went fine
That his son looked like a porcupine.

After two lectures in utter nonchalance
He had answered the telephone askance
But then felt it was a blessing benign
To have a son like a porcupine.

He came to visit his progeny
From behind the doors of the nursery
And pondered-how could he at all combine
His professional career with this porcupine.

And this new responsibility
Made him lose his youthful buoancy
For a while he wondered whether laws divine
Made it not simpler to raise a porcupine.

Five months have passed since those days of consternation
The motive has been changing to one of elation
So that by now he can hardly define
The delight he derives from his porcupine.

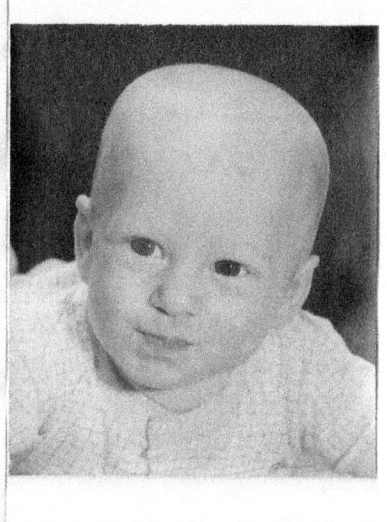

The most important event of our stay at Lehigh was of course the birth of our first child and son, Michael. Michael's arrival on December 28, 1953 was an unmitigated joy for both of us and Marion wrote a beautiful poem for the occasion.

Bethlehem is not too far from New York City and both Marion and I had some relatives there that we occasionally visited. On some of these trips we also had the opportunity to go to museums and the opera. Philadelphia was also fairly close and we occasionally visited there. One particular event that Marion never could forget was our visiting a museum when she was almost ready to give birth and standing in front of a painting of Prometheus being punished for stealing the fire from the Gods. I prefer to remember a dinner that day at Bookbinder's, just before Michael made his entry into the world. Bookbinder's was a fancy fish restaurant where patrons were given a bib before eating. Michael was a beautiful child, and as he grew a little older he had the same blond hair I had when I was a boy. I can still recall a story my parents told me. On a hot summer day we were walking on the Ring in Vienna when a woman stopped her *droschke*, a horse-drawn vehicle. She got out and said to my parents, "I only stopped to admire you son's *Schaumrolle*," (my light blond hair which was combed in a roll).

I believe that I was genuinely appreciated at Lehigh University and if I had so desired, I could have become one of the old tenured professors who loved the early American culture, renovated an old farmhouse, and settled in Pennsylvania. But Marion and I did not feel at home in the Eastern establishment and never expected to settle there, Moreover, Marion experienced periods of depression and I was not able to help her. Except for Marion's unhappiness, I think Lehigh was good to me and for me, and most importantly gave me the opportunity and support to write the heat transfer book, which became a cornerstone of our life.

CHAPTER 16—*PRINCIPLES OF HEAT TRANSFER*, 1958

The area of heat transfer had been my primary focus ever since I graduated from Berkeley and to write a textbook on heat transfer became an obsession with me after I began teaching such a course at Lehigh. This obsession blinded me to the level of depression and unhappiness that Marion experienced at the time. Undoubtedly, my insensitivity to Marion's condition was at least in part responsible for her seeking consolation in an affair with one of our friends, which almost resulted in a breakup of our marriage. However, Marion helped me faithfully and without complaint, typing the manuscript for my book on an old manual typewriter. Sometimes she had tears running down her cheeks as she was typing repeated drafts of the manuscript on a Corona typewriter with a carbon copy without really understanding the text. After three years of unmitigated work with many sleepless nights, my manuscript for *Principles of Heat Transfer* was finally completed. I asked my good friend Adolf Gruenbaum to recommend a good publisher. He suggested Addison-Wesley, who published scientific books of high quality. I sent the manuscript to the headquarters of the publishers, and they responded by saying "We like your book, but the style in which you have written is too colloquial." I had used in my book a direct approach in an effort to directly speak to and engage the reader. For example, instead of the usual mathematical idiom of "one substitutes x in the equation for y in order...," my writing style was, "in order to determine the value of y, you substitute x into the equation..." The publishers said they would consider the book provided I accepted a co-author, and they suggested I contact Professor William Kays at Stanford University. However, I felt confident that students would like the style of my writing and refused to accept a co-author. In retrospect, of course, I realize that the publishers simply wanted to have the name of a person who was well-known in the field and from a prestigious university on the book in order to increase its sales. But at the time I was desolate and really did not know where to turn next.

However, fate intervened. I had submitted an article for a meeting of the American Society of Mechanical Engineers in New York and was asked to make an oral presentation. I arrived in New York by train rather late the night before my talk and went to the hotel where I had made a reservation. However, when I came to the registration desk, the clerk could not find a reservation for me and apologetically said, "This hotel is full, but if you go down the street the hotel next door may have a vacancy." At that point, a man who had just registered and stood next to me said, "I have a large room with two separate beds. Would you like to share the room with me?" Of course I accepted his offer and as we climbed the steps to his room, he asked "What are you carrying under your arm?" I told him that it was a manuscript for a new book on heat transfer. The man then introduced himself as Ken Gromlik, the editor of a small startup publishing company called International Textbook Company. He then asked if I would permit him to take a look at the manuscript, to which of course I happily agreed.

It had been a long tough day for me and even though Ken kept the light on over a small table in the corner while he was reading, I soon fell asleep. When I awoke the next morning, Ken seemed a bit groggy, but asked me to go for breakfast with him. Then he said, "I think your book is outstanding and if you are willing to sign a contract with us, we would be happy to publish it." I was overjoyed because after refusing to accept a co-author I did not expect to find a publisher this quickly. Little did I know that mine may have been the company's first full-scale textbook because the International Textbook Company had previously been mostly involved in correspondence courses. I signed the contract on March 14, 1957 and we immediately began our work on completing details, such as getting permission for certain figures and improving the art to get the book ready for publication. Nine months later I held the first edition of my book in my hands. It was dated January 1958.

Looking back, this night was probably the turning point in my life, both professionally and personally. Although the publisher did not have any sales people and relied mostly on advertising by mail, the book became an instant success. It was the first comprehensive textbook in the field of heat transfer and many universities had began to realize the importance of the subject and had started to teach heat transfer courses specifically designed for mechanical engineers. At Berkeley, we had used a set of mimeographed notes compiled by Professor L.M.K. Boelter in lieu of a text in the heat transfer course. These notes were very helpful to me in writing my book, but they lacked organization and often had only equations without text. The only competing text was a book authored by McAdams, a chemical engineering professor at MIT. But McAdams' text was mostly a compilation of empirical data without a coordinated method of "teaching."

My book assumed that students would have taken a course in thermodynamics as a prerequisite. In the opening paragraphs, I explained that thermodynamics, despite the fact that "dynamics" in Greek means movement, actually assumed transitions from one steady-state condition to another, whereas engineering design required the prediction of the rate of heat transfer. In other words, time was not a variable in the traditional thermodynamics courses, while it was the key in the engineering design described in my book.

At the time I began writing the book, it was generally known that heat can be transferred by three different modes: conduction, radiation, and convection. In most engineering situations however, heat is transferred not by any one, but by several of these modes simultaneously. The novelty and methodology of my book rested on the recognition that a general method for handling heat transfer problems could be based on a thermal circuit analysis. This idea of the thermal circuit was based on the similarities between equations governing the flow of heat and the flow of electric current. Using the analogy between electrical and thermal systems, the book introduced the thermal circuit concept. Interpreting heat transfer problems as a thermal circuit was a tool that engineering students, as well as professional engineers found very useful. Once one acquired a physical understanding of the analogy, it was only a matter of quantitatively evaluating the various driving forces or potentials, which in heat transfer is a temperature difference, and the thermal resistances based upon laws of heat transfer, as well as the capacitance of non-steady state conditions which often are encountered. The treatment of each of the modes of heat transfer—conduction, convection, and radiation—was fairly standard. However, the simplification of introducing an equivalent thermal resistance for radiation was new and made it possible to combine all of the modes of heat transfer under a single umbrella. A favorable review introduced my book to a wide audience.

226 HEAT TRANSFER BY RADIATION

collecting surface. However, the cost of power production would be approximately three times the present power cost by more conventional means.

Where high temperatures are desired, some form of concentrating collector must be used. Many different combinations of mirrors and lenses have been proposed, and some arrangements have actually been used for solar engines and furnaces. The advantage of concentration is that the area from which heat losses occur is less than the area receiving radiant energy, and the heat losses are proportionately reduced with an increase in equilibrium temperature. Figure 5-37 shows a conical-mirror arrange-

Fig. 5-37. Solar motor. (Courtesy C. F. Holder, Ref. 27)

ment used, in 1900, by A. G. Eneas (27) to generate steam at 150 psi. With a mirror surface of about 700 sq ft, a steam engine was driven for which an output of 10 horsepower (hp) was claimed.

Another novel feature of the book was that the topic of solar energy conversion was introduced and treated in some detail. One chapter in the book discussed the thermal conversion of solar radiation into process heat and electric power. The page from the book shows the first solar engine built at the end of the 19th century. The basic approach of concentrating solar radiation in order to achieve higher temperature is still being used in many of the current solar systems with great success. The chapter also introduced the potential of photovoltaic energy conversion to produce electricity, as well as the use of flat plate solar radiation collectors to heat water. These types of solar energy systems are today one of the most successful applications because they utilize the incoming radiation all year round and therefore are economically viable. One of the problems in the book also dealt with the possibility of global warming, which at that time was not a part of the environmental vocabulary. Some basic aspects of nuclear power was also included.

The timing of my book's publication was very propitious because teachers in the field were all looking for a text. Despite the publicity limitations of the International Textbook Company, which had no staff to complete with giant book companies such as McGraw-Hill and Wiley, the book found wide acceptance and eventually as many as 160 universities adopted it as a text. Many of the teachers using my book began to interact with me, mostly by mail, making suggestions for improvements, correcting mistakes, and offering additional problems to be included in a future edition of the book. For my own self, I felt that the success of the book demonstrated my professional competence and hoped it could substitute for my lack of having completed a PhD at Princeton.

To be honest, I did not realize just how original my book was. But other publishers saw that there was an opportunity to penetrate the field with similar books and soon began to look for authors willing to write competing texts. Lured by the promise of a hefty advance, several authors began writing competing texts. All of these texts were based on the same basic approach that I had pioneered. In fact, the first competing text that appeared was by Professor Holman and was so similar to mine that it was called "the little Kreith." The International Textbook Company considered suing for plagiarism, but decided that such a suit would be expensive and might not be successful. Ken therefore asked me to begin work on a second edition, and by improving the book, continue sustaining the large number of adoptions. I accepted his suggestions and began work on the second edition, which was completed in 1965. *Principles of Heat Transfer* was probably the most important contribution to the field

of engineering that I made. There have been 7 editions so far, four of them with Dr. Mark S. Bohn as co-author. Today, 55 years after the first edition appeared, a new publisher, Cengage, Inc. showed their high respect for the book by initiating the preparation of an eighth edition with Professor Raj M. Manglik as co-author.

CHAPTER 17—MOVING TO BOULDER, UNIVERSITY OF COLORADO, 1959

Both Marion's and my parents lived in California and every summer we drove from Pennsylvania to California in order to be with our parents and have them get to know their grandson, Michael. On the way, we passed through Colorado and I fell in love again with the mountains, with the snow, and the fresh air. I knew very little about the University of Colorado, except that a Professor Cohen had established an Honors Program there, but I stopped every time we passed through Boulder and left my curriculum vitae with the College of Engineering hoping that someday an opening for me would turn up. Unfortunately, in the Mechanical Engineering Department, a Professor Ben Spurlock claimed the title of "Mr. Heat Transfer" and did not want any competition in his department. Ben had been writing a heat transfer textbook for the past 10 or 12 years, but apparently had never progressed beyond the second chapter. However, to my great surprise, in the summer of 1958 I received a letter from K.D. Wood, the Chair of the Aeronautical Engineering Department asking whether or not I might be interested in a position as full professor. He told me that any appointment to the University would require a two year probationary period, after which tenure would be granted. I was only too happy to be able to come to Colorado and get a promotion, and had enough confidence in myself not to be seriously concerned about giving up my tenure at Lehigh. Professor Wood asked if I could join the department in mid-year, and also bring my research equipment with me. I requested to be relieved of my contract from Lehigh and the university administration graciously agreed and also allowed me to take my research equipment along.

Our move from Bethlehem to Boulder in January of 1959 was uneventful. The last few weeks in Bethlehem had put an enormous strain on the relationship between Marion and me and the prospect of a new start in a beautiful part of the country provided a spark of hope for the future. When we arrived in Boulder, we found a small town of about 10,000 people with a single Safeway supermarket, a movie theater at the edge of town, many still-unpaved streets, and only one eating establishment, located in the local drugstore. The town was dry, except for a hotel by the name of Harvest House, which had carved a space for itself barely inside of the city, but legally in the county and was allowed to serve liquor. But the prohibition on sale of alcoholic beverages did not deprive the faculty of their martinis. Every week one of the faculty members took orders for a liquor run to Harry Hoffman (a liquor store) in Denver. I am not too surprised that these orders were always delivered on time.

The university was similar to that of other agricultural state institutions of the time. The campus was beautiful, located in view of the mountains on one side and the plains on the other. The student body, consisting mainly of undergraduates, was about 10,000. Most of them were Colorado residents because there was academically little to attract out-of-state students. Skiing was still in its infancy and the football team was barely on the map. The faculty was, by and large, undistinguished. Emphasis at the university was mostly on undergraduate teaching and faculty members could attain tenure without a PhD and little or no research accomplishments. A normal teaching load was three lecture classes per semester, each worth three units. Each professor was expected to grade homework and tests for those classes, which consisted of between 30 and 40 students. This compares with a normal load of one class per semester today. It is not surprising that this load left little time for research and scholarly writing. Sabbatical leaves were not part of the university culture, although one could apply for a leave with pay for special projects. This had to be approved by a faculty committee as well as the department head. Department headships were permanent appointments.

The tenured faculty in the Department of Aeronautical Engineering consisted of its head, Dr. K.D. Wood, and one professor, Mr. Seibert, both of them in their early 60s. The rest of the faculty members were untenured lecturers, some of whom were still working on advanced degrees at other universities. A day or two after I arrived, K.D. called me into his office and said, "the department expects an accreditation visit within a year and your presence is expected to provide two needed ingredients for passing accreditation: One is sponsored research, and the other is a publication record, primarily your newly published text, *Principles of Heat Transfer*." There were no graduate courses in the department, but the laboratory, although fairly old, was kept in relatively good shape by an old technician who had once been a miner.

Approximately a year after I arrived, the status of the department was examined by a national engineering accreditation team and barely received accreditation to award Bachelor's degrees for the next five years. I was told that my previous associations with the California Institute of Technology and Princeton University helped enormously in the evaluation process of the department's curriculum.

We were living in a house that we had rented on Baseline, which was a major thoroughfare with noisy traffic and no view of the mountains. However, within days of arriving, I began to look for a place where I could build my own house, above the city and with a view of the mountains. A realtor showed me around the neighborhood and we came to a small street with only five houses called Sierra Drive in a newly developed subdivision, which had a lot overlooking Boulder. The place reminded me of our first apartment in Berkeley, from which we could see the city of San Francisco, albeit only from the bathroom if one stood on the bathtub. The lot in Boulder was looking down on the city on one side, with beautiful views to the east and west. At the time I gave very little thought to the problems of building a house on this steep lot. I had the feeling that living on a hill from which I could see the city lights at night, close to an old summer recreation village called Chautauqua, was a compensation for not having been offered a tenured position at Berkeley, where I had dreamed of finding a place overlooking San Francisco. Here I was not only offered the view of the city, but I could also see a mesa with four trees silhouetted in the east and a beautiful mountain backdrop to the west.

A herd of deer and a fox were occasional visitors and the property was close enough to the university to walk or bicycle, yet far enough not to be in the city. It was a dream comes true, and after I bought the lot, I repeatedly came to it and just marveled at my good fortune. Our house was finally built in 1967. A few weeks ago Marion asked me, "What was the best decision you ever made?" and I said, "Building our house on Sierra Drive."

In addition to my teaching and research activities in aeronautical engineering, I became interested in the Honors Program in the College of Arts and Sciences, which at that time was limited to students in that college. The program was under the direction of Wally Wier, a professor of philosophy. Wally was an unusual man. He was good looking, well read, expressed himself beautifully, and was able to establish excellent interactions with his students. Wally enjoyed running the Honors Program where only students with high grade-point averages were permitted to participate and small classes were based on discussion rather than lectures. Wally was a fascinating man and we became good friends. The students adored Wally and he provided stimulation in their academic endeavors. Wally also enjoyed spending time outside of classes with students, particularly young women, and it was rumored that their interactions were often not merely intellectual. But none of the women ever complained.

Wally, who had not had any previous contact with technology, became very interested in engineering. He enjoyed working with me and proposed setting up a joint Honors Program that would bring students from engineering and the arts and sciences into contact. I was able to receive permission from the Engineering Dean, Eckel, to organize an Honors Program for engineering students in close cooperation with the existing arts and science program. This activity was not only one of the highlights of my early years at CU, but later on also earned me a tenured position in the Department of Mechanical Engineering. Marion began to continue her education and enrolled for in the French Department. She and Wally's wife Janet became close friends and in 1964 both of them received their BA degrees at the same graduation ceremony. Given her lack of formal education, this was a truly great achievement for Marion. She went on to obtain a graduate Degree in French in 1972 and taught French while enrolled in graduate school. Later on she taught courses of French Literature in Translation in the Honor's Program.

As has been so often true in my life, I was too involved in professional activities and also taken up by the personal problems within my family to see dark clouds on the horizon. In an unguarded moment, K.D. Woods' wife referred to one of the faculty wives as a "painted Jewess" and I also heard the word "kike" used in the office when no one was aware of my presence. But I was nevertheless flabbergasted when, in May of 1959, soon after the evaluation team had completed its evaluation and given approval to grant degrees to the department, K.D. Wood came to my office and informed me that "In my opinion you seem really more interested in science than engineering and I will therefore not renew your contract beyond the second year." The news came at a particularly bad time because the strain in my relationship with Marion had reached a breaking point and we had decided to separate.

CHAPTER 18—NOMADS THEATER, 1959–1961

"If you become part of the suffering, you'll be entirely lost"
Anneliese (Anne) Frank:The Diary Of A Young Girl.

In the fall of 1959, a new and totally unexpected element entered my life. The secretary in the aeronautical engineering department in which I was teaching took an interest in the way K.D. Wood was talking about me. She often overheard conversations about K.D.'s machinations and told me surreptitiously of things going on behind my back. At first I thought it was simply that she felt an injustice was being done, but later on I realized that her interest went beyond that of a secretary observing the machinations of academia and had a more personal element to it. The secretary's name was Claire S. and she was the wife of a professor of law at the university. She was an attractive, slightly built woman with a beautiful face, but she had a permanent limp due to having been afflicted with polio as a child. Despite this impediment, however, Claire had a great deal of self-confidence and was somewhat of a social butterfly. As I discovered later, she was also an exceedingly talented actress.

PLAYBILL

THE MUSIC BOX

THE DIARY OF ANNE FRANK

Claire and several other aspiring actors and actresses had contributed some money to found a provincial theater called the Nomads. A Boulder couple interested in fostering local theater had purchased an old Quonset hut at the edge of the city which served as the venue for this local amateur theater company. The company had put on some successful plays before, but in the fall of 1959, the Board of Directors decided to obtain the rights to a recent success on Broadway entitled the Diary of Anne Frank. I am sure that by now everybody knows the name of Anne Frank. She had written a diary while taking refuge in the upstairs apartment of a Dutch couple in Holland. Her family, consisting of her parents, her sister, and herself, in addition to some friends and their son, and another couple lived in the attic of the house belonging to this

Dutch couple for two-and-a-half years. About a year before the end of the war, a workman became suspicious and informed the Gestapo of the presence of people living in the house. The Gestapo arrested the entire group and they were sent to a concentration camp where Anne died of typhoid a few weeks before the end of the war. The only survivor was Anne's father, who discovered the diary in the attic after the war. The diary was published as a book and subsequently made into a play. Obviously, the play had enormous emotional attraction for me, and Claire suggested that I should try out for a part, since I was so interested in the story. I had never before thought of becoming involved in the theater, but thanks to Claire's encouragement, I went to the audition for the play at the Nomads. To my big surprise, I won the part of the Dutchman who had hidden the Frank family during the war before their discovery. The main parts in the play went to Henry Erman, a professor of political science at the university who played Anne's father; and to Claire, who obtained the part of Anne. Anne's story has lived on and on May 10, 2014, 55 years later, the Boulder Chorale performed an oratorio based on her life by James Whitbourn at a local church. Both Marion and I attended.

I still recall the excitement when the play opened at the Nomads to rave reviews, but in all this complicated and emotionally stressful part of my life, the interactions between me and Marion were labored and resulted in a lot of grinding of teeth on both of our parts. I do not believe that our relationship would have survived if it had not been for the arrival of our first daughter, Marsha, on November 4, 1959. She was the most beautiful little baby girl and I instantly fell deeply in love with her and through her established a renewed relationship with Marion. We were still living in our little house at 1911 Mariposa, but now began thinking about the possibility of building our dream house overlooking Boulder on Sierra Drive.

My involvement in the Nomads Theater, which had started so unexpectedly, continued for many years. Although I was a mediocre actor who could only handle small parts, the experience of theater life and the interactions with people outside academia was exhilarating. I will never forget the thrill of the moment when the curtain opens on opening night. I still have many of the playbills, as well as some of the scripts from several of the productions in which I participated. The more memorable of those were The Life of Dylan Thomas, The Crucible, The Story of Marasad, and The Dybbuk. The directors at the Nomads seemed to anticipate my tryouts, because I was usually typecast as a religious figure, such as the Reverend in The Crucible and Rabbi Mendel in The Dybbuk. I still remember my audition for the Marasad play when I put on my ski jacket the wrong way so that my hands were bound as in a straight jacket. I did this because I knew that the character that I was trying out for was a mad priest in a straight jacket. My participation in The Dybbuk resulted in a lifelong friendship with another man who tried out for Rabbi Shimshon. His name was Herman Senft and he was married to Edith, a woman a year or two younger than I. Both were refugees, Herman from East Germany and Edith from Vienna. Edith had survived the war in Shanghai and both of us shared common memories of Vienna during the "Good Old Days."

CHAPTER 19—NATIONAL BUREAU OF STANDARDS, 1959-1961

In 1954, President Eisenhower had inaugurated the new R&D Building at the National Bureau of Standards (NBS) in Boulder, Colorado. One of the key areas of interest of the Bureau at the time was cryogenic engineering (cryogenic means ultra-cold). In the summer of 1959, I was asked to become a consultant to NBS. Given my experiencing in heat transfer and the need for the U.S. Air Force to develop thermal insulation materials suitable for tanks holding liquid oxygen and liquid nitrogen on rockets for the lunar exploration program, engineers at the Bureau thought I could be of assistance. I was asked to work with other engineers at the Bureau on the development of low conductivity thermal insulating materials for cryogenic fuel tanks suitable for spacecraft. Before accepting this position as a consultant, I requested clarification regarding the classification of the project and was told that it was an unclassified project—that is, the project that did not require any security clearance. I worked on this project for over a year and published two papers which were presented at national conferences. At the end of 1960, I was told that it was normal procedure to fill out an application form for the National Bureau of Standards, which I did on June 1, 1961. A few weeks later, my supervisor, Dr. Kropschot, who was also an adjunct professor at the University of Colorado, came to my office and said that he would permit me to present our joint new article, but that I would not be permitted to enter the Bureau any longer. When I asked him for the reason, he said that the FBI had provided information with certain charges that the Director of the Bureau had reviewed and, based on the review, had decided to deny me entrance to the Bureau. Subsequently, I obtained a copy of the letter which had been sent from an unknown source to the Director of the Bureau repeating the same old charges that had been leveled against me in 1951, and which upon the interview and interrogation were found to be unfounded. I went to the Director of the NBS and explained to him that in order to avoid any problems when I started work at the NBS, I had requested that I would only be assigned tasks that were unclassified. The Director responded by saying that he was ultra-careful and that, although my work did not require any security clearance, my presence at the Bureau potentially gave me access to material that might be classified. The Director then told me "If you want to continue work at the NBS, you will have to undergo a new security investigation. But this should be no problem if you have nothing to hide." I responded by telling the Director that my work was unclassified and

I considered it demeaning to have my loyalty questioned without there being any rationale for doing so. I then walked out of his office and never again returned to the National Bureau of Standards. Strangely enough, the Bureau published the results of my work as part of NBS work to develop a cryogenic insulation for liquid propellant rockets, which were subsequently used to send a man to the moon.

CHAPTER 20—CU MECHANICAL ENGINEERING DEPARTMENT, 1961–1967

After the termination of my two-year appointment in the aeronautical engineering department in January of 1961, I was invited by Professor Robert Williams to transfer into the mechanical engineering department where I found a more benevolent environment that allowed me to put emphasis on my professional career. The department appreciated my work in the Honors Program and I was given tenure after one year. Also Marion and I began to find common ground on which to build a more mature relationship. This process of finding a foundation for our marriage was helped enormously by the arrival of our wonderful second daughter, Judy. I remember that she was a feisty baby from the very start of her life. Today she is a beloved dance teacher in our community and has taken a personal interest in bringing Marion's survival in Cuba to a wide audience.

In early 1961, Dean Eckels, who had been at the helm of the college for many years, announced his retirement. Eckels was an old-fashioned engineer interested mainly in undergraduate education and placed very little emphasis on research. This attitude was changing under the direction of the new president, Quigg Newton, and the newly appointed provost, Oswald Tippo, both of whom wanted the College of Engineering to enter the 20th century by doing forward-looking research in the college. The only department at the time that seemed to share this view of the future was electrical engineering. The head of that department, Willis Worcester, courted me and asked for my support to nominate him for the deanship. For some reason or another, at the time I felt that an outside person would be able to change the culture of the college better than someone from the inside. That was probably one of the many stupid decisions I made in my life, because Willis supported me and my ideas for engineering education. But the administration wanted to have an outside nominee, and Professor Klaus Timmerhaus of chemical engineering threw the name of his friend Dr. Max Peters, a professor chemical engineering from the University of Illinois, into the ring. Given the limited support in the state for higher education and the lack of research-oriented faculty in the college, there were apparently not many good

---

candidates interested in the deanship. Hence, Max Peters was invited to interview for the position. I first met him for a 10-minute interview in his hotel room. Max was a short and stocky man with a crew cut and a raspy, clipped way of speaking. After a brief introduction I introduced the one item of prime importance to me. I told him about the success of the Honors Program in engineering that I had initiated and suggested that it could be made into an engineering science department, which at that time attracted many outstanding students in other universities. I recall Max's reply, "This doesn't make me jump with joy." I should have realized at this point that there were vast differences in our respective ideas for the future of engineering at CU.

Some time ago, Professor Irving Weiss, a colleague from Lehigh who had also moved to CU, told me some interesting facts about the family background of Professor Timmerhaus. Timmerhaus Sr. was an expert in communication technology, a technology of critical importance to modern warfare. He came originally from Germany and had been in touch with the German government for some time before the outbreak of the Second World War. When he was offered in the spring of 1939 a key position in the German war machine, he decided to leave the U.S. and return to Germany. The family had booked passage on a ship to Europe soon after Hitler invaded Poland and the only thing that prevented the family's return to Germany was the U.S. declaration of war after Pearl Harbor. Klaus never expressed his feeling about Hitler or anti-Semitism to me, but when I met him the first time he reminded me of an SS Officer who had just taken off his uniform. Not surprisingly, Max Peters appointed him Associate Dean soon after his arrival and Klaus remained his staunch supporter. Max and Klaus shared a passion for athletics and I am told that both did a two-mile run every day before breakfast.

When Max settled into the position of Dean at the college and I had an opportunity to interact with him more directly, it quickly became fairly obvious that there was a clash of cultures and personalities. Max had been a sergeant in the Mountain Division as a ski trooper and he treated people working for him as a sergeant would treat his soldiers. He wanted to see sponsored research in engineering, directed towards traditional engineering topics of interest to oil and gas industries. My interest in interdisciplinary studies, research in the interaction between biology and engineering, development of solar energy, and studies on the environmental impact of energy, were topics of no interest to Max. I felt, however, that I had achieved a fairly important position in the college and did not realize that any opposition to Max's ideas was doomed to failure. This began a battle between Max and me that was fought with very unequal weapons. Unfortunately, as a result of my background and personal experience of a dictatorship, I considered it virtuous to fight against the Dean's autocratic leadership. How foolish can you be?!

Although I did not receive the new Dean's support for supervising the Honors

Program in engineering, I simply plowed ahead and continued to interact with my counterpart in the College of Arts and Sciences, Professor Wally Wier. As it turned out, our ideas of interdisciplinary education were quite similar and did appeal to many outstanding students who continued to participate in the Honors Program. One of the most innovative courses in this program was a course entitled "Science and the Modern World" that Wally and I co-taught every fall semester. In that course, we attempted to show students how modern science and technology had affected the development of Western culture and what the potential shortcomings were of these developments, such as pollution, crowded cities, depletion of resources, dependence on the automobile, and population explosion. I learned a lot from this course and continue my interest in interdisciplinary education to this day. In the latter part of the book I will describe how I expanded this interdisciplinary approach into a course on sustainable energy.

My situation in the Mechanical Engineering unfortunately deteriorated when Max announced the appointment of a permanent chair for the department to replace Professor Williams in 1962. He had met the candidate, Dr. Frank Essenberg, at the railroad station in Chicago and decided on the spot that the man was the ideal person to lead the department. Frank had originally studied law and passed the bar, but after a few years of practice decided that he wanted to become an engineer. He then obtained a PhD in Theoretical Mechanics at the University of Illinois, where he held a temporary faculty position at the time Max met him. Max did not consult with anybody in the department, and Frank Essenberg appeared on campus as the Chair of Mechanical Engineering at the beginning of the academic year.

Frank and I disliked each other from the moment we laid eyes on one another. He was an obese man with glasses and had a slimy handshake with fingers that felt as though you were touching five eels. He had a loud voice and held the opinion that all of mechanical engineering was derived from theoretical mechanics, and that heat transfer and thermodynamics were largely empirical and therefore not a significant part of the curriculum. When I asked him what he thought of the first and second law of thermodynamics, he replied, "Those are only based on empirical observations." Although Essenberg had very little understanding and interest in the broader aspects of mechanical engineering, he and Max Peters formed a close personal relationship. Some years later at the funeral for Essenberg, Max Peters said little about Essenberg's professional accomplishments, but told the mourners that Frank made excellent dry martinis, which the dean apparently enjoyed drinking with him.

The first major clash between Max and me occurred about a year after he arrived. I was serving at the time on a committee to look for graduate students that could also become research assistants to some of the newly appointed faculty that wanted to start research programs. As part of my responsibilities, I asked for as much information as possible from each of the applicants. The dean assigned the task of responding to our recommendations to his assistant dean, Chuck Dedi. Chuck had been a highly decorated Colonel in the U.S. Army. When he entered a room he would stop, stand at attention and look at you with a stern glance. One time he came to one of our committee meetings with a photograph of a group of spectators at a football match. One of the people in that photograph had applied for an assistantship in the Engineering College and was clearly Asian. Dean Dedi pointed at that person and told the committee that, "Our dean would like more good American boys in the college, and we do not want another gook." I was stunned by the way the dean wanted to make his selection and considered this clearly to be a policy of racism to which I was violently opposed. I began to search for other people in the college that had similar views and found among others support from Ken Martin, the only Black member of the engineering college faculty. Ken was a temporary lecturer in the Department of Engineering English, which the dean had established because he felt that the regular Department of English was too liberal for engineering students. Ken and I found a few more faculty members willing to publicly oppose the racist policies of the dean and asked for a committee investigation. This investigation resulted in an open hearing organized by the Vice-Chancellor. In the course of the hearing, everyone in the room was asked whether they had personal experience with racist attitudes from the dean. To this day, I feel ashamed for not having spoken up clearly. I did not have any personal knowledge of the dean having made racially motivated decisions, other than, of course, having appointed an assistant dean who carried his message. I should have spoken up more forcefully and explained some of the indirect observations that were indicative of racism—if not of direct racist policies that discouraged diversity in the faculty and students. I am not sure whether my background with a dictatorship made me momentarily afraid of the dean who had the demeanor of a dictator, or whether I was hoping that by not speaking up I could establish better relations with him. But to my everlasting shame I kept quiet.

The outcome of the investigation was that Assistant Dean Chuck Dedi was dismissed. We were told that Mr. Dedi was a heavy drinker, had recently lost his wife, and had acted on his own prejudice without the approval of the dean, which, according to the words Dedi used, was obviously false. But by dismissing Chuck Dedi, the administration had found a scapegoat, cleared Max Peters of any personal involvement in racist policies, and thus ended the incident officially. But, as far as the relationship between Max and myself was concerned, it obviously only intensified our antagonism. From that time on, it became clear that Max wanted to find a way to force me to resign by making my professional life unbearable. The Black faculty member, Mr. Martin, understood the situation better than I did and immediately resigned. I relied on my tenured position, which provided job security and income for my family. One of the main reasons for this decision was that ever since I had been declared a national security risk, I suffered from a lingering fear that I could not find other employment due to the cloud of suspicion created by having once been denied security clearance. Moreover, I think I loved Colorado and was not willing to forgo the beautiful setting we had created outside the academic environment.

I was not the only faculty member who accused the dean of racism. Max Peters hired Professor Mahinder Uberoi in 1963 to be the Chair of the Aeronautical Engineering Department. Mahinder was a tall imposing man with pitch-black hair and an aristocratic bearing. He had an excellent reputation as a subsonic aeronautical engineer at the University of Michigan and the department seemed to improve under his leadership. But something happened to sour their relationship and Dean Peters suddenly removed him from the chairmanship and appointed Frank Essenberg as Chair, although Frank had no experience in aeronautics, to replace him.

Uberoi accused the dean of having removed him from the chairmanship because he was of Indian background. He claimed that he was brought to the University as a token minority faculty member and had done a good job in guiding the department before his dismissal. He then filed suit against the dean and the University for Racial Discrimination. The legal maneuvering continued for many years without a settlement, but none of the details of the suit were ever divulged. I do know, however, that Dr. Uberoi never again set foot on the University campus, but continued to receive his full salary until he was dismissed from his tenured status in 2000 by the Board of Regents, many years after Dean Peters had been forced to resign.

Uberoi shared his love of British sports cars with me and kept a small white Alpine convertible, similar to one I had, in his backyard for many years. But after the dismissal from his Chair at the University, he withdrew from any social contact with his previous colleagues. Mahinder passed away in 2006. Marion and I attended his funeral service in Golden where we learned that he had been a Sikh. The Sikhs had suffered various acts of suppression over the years and although they constitute less than 2% of India's population, they had made significant contributions in all walks of life in India, similar to the Jews of Europe. The service in Golden was quite moving, but only one other faculty member attended. Mahinder left all his belongings to the Uberoi Foundation for

Dean of Engineering Adds Name to List Of CU Resignations

BOULDER—Max S. Peters, dean of the University of Colorado College of Engineering and Applied Science for 15 years, has resigned to resume teaching.

The resignation will be effective as soon as a replacement is named, but no later than June 30, 1978.

Peters, 56, didn't do well in a recent faculty evaluation of his performance. An engineering school source said teachers have been unhappy about reorganization of the academic unit, "variable support"

Religious Studies, which was formed by some of his friends in his honor.

In 1977, Dean Max Peters was forced to resign after a faculty evaluation of his performance showed that the professors in the college were very dissatisfied with his leadership. The University then turned to the son of a former CU physics professor, Dr. Pietenpol, to take over the deanship. He was a nondescript middle aged man with a sonorous voice who looked more like a high school teacher then a dean. But he seemed like a reasonable person to me. In view of President Carter's strong support of solar and renewable energy, I decided to try once more to interest the college in solar energy and proposed a solar program to the University, citing the favorable location of Boulder, the nearness of SERI, the interest of members at SERI to gain advanced degrees, and the great potential for research between the two organizations. I suggested that the University should offer a degree in Solar and Renewable Energy Engineering, similar to that offered at two other first-rate universities. In an interview conducted by a reporter from the *Colorado Daily,* Dean Pietenpol said "Kreith's proposal doesn't make much sense. Solar engineering is only one part of the total engineering project picture. If we offer a program in solar engineering, then we have to offer one in nuclear engineering, coal engineering, and oil engineering. We will not change to offer a degree in solar engineering."

tuesday- oct. 30, 1979 - the colorado daily - page 8

Prof. urges establishment of CU solar engineering program

By CYNTHIA SANTANA
For the Colorado Daily

Not only is there a shortage of energy today, but there is also a shortage of solar engineers with which to combat the problem, according to Dr. Frank Kreith, chief of the Thermal Conversion Branch at the Solar Energy Research Institute in Golden.

To help alleviate this problem, Kreith wants to build a solar engineering program here at the University.

"Solar designs require a high level of engineering know-how," Kreith said. "Poorly engineered systems and poor installations have plagued the solar industry. There is not a sufficient number of solar engineering programs—not a single one exists on any campus in the United States. There are currently only solar technology programs."

He made reference to Science magazine, which reported this month that between 1972 and 1977 the number of U.S. citizens who obtained PhDs in engineering decreased from 2,329 to 1,507, a 35-percent reduction.

In the same article, by Roy A. Young, chancellor at the University of Nebraska, Kreith underlined the following statement: "Shortages will become even more critical as major synfuel production efforts are launched and solar research and development activities are increased."

"Solar experience is generally quite limited," Kreith said.

Kreith, a native Austrian regarded as one of the foremost authorities on solar engineering, has been on leave of absence from the University since 1977 working as one of the original members of the management teams at SERI.

Several proposals to initiate a solar engineering program at the University have been offered by Kreith to Engineering Dean William Pietenpol. "The University of Colorado would be ideal with the location of SERI, the interest of members of SERI to gain advanced degrees from CU, and the support of thesis work between SERI and the University," he said.

As of yet, Kreith said, he has not received any indication on the status of his proposal, but speculated that one reason the University might be hesitant to establish a solar program is because "traditionally the University has received a great deal of support from the oil industries." He added, however, the oil industry has recently shown an interest in solar energy in the reactivation of drained oil fields utilizing steam.

"I have a loyalty to CU, and I hope CU will support my desire to build a solar program. If CU will not do it, I will be forced to look elsewhere," he said.

According to Dean Pietenpol, many of Kreith's proposals "don't make very much sense.

"Solar engineering is only one part of the total engineering project picture. If we offer a program in solar engineering, then we have to offer one in nuclear engineering, coal engineering, oil engineering," Pietenpol said.

"At the moment we will not change to offer a degree in solar engineering; it is not planned. We do have solar activities in other departments such as electrical, mechanical and chemical engineering, and we're also working with a grant from SERI," Pietenpol said.

"Solar engineering is only one effort in the problem, and we don't want to overplay the importance of solar power. When it becomes important to educate, we will be a part of the program—we are currently a part of it," he said.

Regardless of how the proposal turns out, Kreith said he will continue to work in the development of solar energy engineering. At SERI he is currently researching methods for reducing consumption of non-renewable energy sources by developing energy conservation and economically viable solar energy systems for industry and domestic use.

"We are working on methods for matching solar conversion technologies to U.S. industrial needs and obtaining a maximum of energy for a minimum capital investment," he said.

"Twenty-seven percent of our energy needs are for heat in industry—this is a large market segment. Former Secretary of Energy Schlesinger made these applications the prime target of our national solar program," Kreith said.

In two weeks, Kreith will attend a workshop with Gov. Lamm and the Solar Energy Advisory Board to set 1980 priorities for the state of Colorado. Kreith would like to see energy effective and energy conservation designs required in all new homes, specifically for domestic hot water systems.

"The conventional gas system is wasteful. It produces an excess of hot water and then uses cold water to reach the desired temperature. This is thermodynamically inefficient because heat is lost when the water is cooled. A solar energy system just tries to get to the desired temperature," Kreith said.

He would also like to initiate a program similiar to one in Portland, Oregon, which requires a house to pass an energy audit before it is well insulated before it can be sold.

Kreith feels there should also be some state legislation, such as tax reductions and low-interest loans, to encourage the integration of solar designs and installation of solar collectors in homes.

On the international level, Kreith believes the use of solar energy can lead to a peaceful future. "Solar energy is available to everybody," he said.

According to Kreith, it would be very easy to transfer technical aid to developing countries to help them establish their own solar industries, utilizing local labor and local materials in order to reduce costs.

He pointed out, however, that a standard of performance has not yet been established which is necessary to sell devices. At the National Bureau of Standards, the Center for Building Technology's Solar Technology Program is currently working to establish and revise guidelines to judge the performance of solar energy systems.

Kreith feels Europe has always been more energy conscious than America. According to Kreith, Israel is making use of the largest amount of solar energy, and Sweden is the most advanced in energy efficiency.

Considering the role of solar energy engineering in the overall energy picture, Kreith believes his field of research "is really trying to fill a vacuum with a new source. The new source has problems and has to be adapted to the needs as they are created," he said.

"American people never had to pay replacement costs in energy. They have lived off of energy savings expecting energy to be plentiful and cheap. However, we are coming to the end of our savings and we must start working to catch up with our needs," he said.

I was disappointed that the new dean was also so shortsighted regarding the potential of sustainable energy, but without the dean's support I realized that no one at CU would take the initiative for a program in renewable energy. Two weeks later, I participated in a workshop with Governor Richard Lamm and his Solar Energy Advisory

Board to set 1980 priorities for the state of Colorado, where I found a good deal of support for my ideas regarding incentives to promote renewable energy in Colorado. Today Colorado is the second most advanced state (after California) in generating the largest percentage of its energy from renewable solar sources. It was not until 20 years later that the University finally decided to offer a degree in renewable energy. In 2000, the University of Colorado and the Solar Energy Institute formed a partnership named the Renewable and Sustainable Energy Institute (RANSEI) that offered a Professional Renewable Energy Certificate Program to provide an in-depth study of renewable and sustainable energy technologies, policies, and business—exactly what I had proposed to Deans Peters and Pietenpol years ago. The College of Engineering renamed one of the laboratories the Center for Renewable Energy and in 2014 the college announced that an entire Sustainability, Energy and Environmental Complex would be built on the East Campus. So at long last I saw my premature hopes for the university fulfilled.

CHAPTER 21—SABBATICAL LEAVE TO FRANCE, 1964–1965

Towards the end of 1963, I realized that the lack of a doctoral degree on my name handicapped my ability to attract graduate students and grants as the university moved from a teaching to a research institution. I did not feel that at this point in my career it would make sense to begin normal academic studies towards a PhD. However, I learned from a friend of mine, Dr. Andrew Charwat, whose family had longstanding associations with the University of Paris, that it would be possible to obtain a doctoral degree from there by doing advanced research for one year. At the University of Paris was an internationally known Heat Transfer expert by the name of Professor Edward Brun. He was also the Director of the well known Laboratoire d' Aerodynamiques. I wrote to him about the possibility of doing advanced research at his laboratory and he replied enthusiastically that he would welcome me at his laboratory if I were to come to Paris. The University of Colorado did not give sabbatical leaves; however, it did grant faculty fellowships for up to one year to do research at another institution. I applied for such a fellowship and was successful in obtaining this grant. Marion, who spoke perfect French from her education in Belgium, was enthusiastic about going to France; but I had to begin learning enough French to be able to function in a French-speaking environment. I had considerable difficulties in learning another foreign language, but did progress to the point where I could read technical papers and speak well enough to be understood in French. One of the wonderful facets of our trip to France was that we traveled by ship on the famous luxury liner, Queen Mary, that was much slower, but also more elegant and exciting than an airplane. Today these beautiful liners are only used for short cruises but hardly for cross-Atlantic transportation.

I wrote to Dr. Brun and proposed that I would continue my research in rotating systems at his laboratory and study the heat transfer between two rotating disks. This assembly was similar to that of modern gas turbines and Professor Brun agreed that it would make a suitable research topic. Thus, in the summer of 1964, Marion and I took our three children to Paris. It was amazing to see all my children learn French very quickly once they began going to French schools; however, having a 4-year old learning a language on the street can create mishaps. One day Judy came home from kindergarten and innocently announced "Tu sais, Maman, elle est putaine la maîtresse" (You know, Mom, the teacher is a slut). I had considerable difficulty understanding French-speaking natives. Moreover, when they realized that I did not understand, they quickly switched to English, which most of them spoke more fluently than I spoke French. The year in France was a wonderful experience for all of us. I managed to get my research done in a timely fashion and was able to submit a thesis to the graduate committee a year later. To my chagrin I discovered that I had forgotten to register for admission to the graduate school. But with a little help from the authority, the registrar relented and allowed me to register "as late." In addition to the main research, it was also necessary under the conditions of the doctor's program to submit a secondary paper which had to be orally defended before an audience of peers. For the secondary paper, I chose "Problemes de Rayonnement Lies aux Vols Spatiaux." Below is a typical page from my thesis demonstrating the mathematical equations that had to be written by hand. The analytic complexity was a challenging mental exercise and the results were published in French and English peer-reviewed journals. The work earned me a Doctorate, but I felt that the effort was wasted because the work did not advance my personal life goals one iota. The work was original in that it had not been done before, but anyone with an adequate background in partial differential equations could have done it and while the experimental work was novel and my own, it was not on the level of my research at JPL where I had skilled technicians to help me. I vowed never again to get involved with research and publications for the sake of academic convention, but seek out topics with a significant focus that would contribute to my vision of a better world.

In addition to my research at the laboratory, Professor Brun introduced me to the publisher of French textbooks. I told the publisher about my text in heat transfer and he suggested that I send a copy of the book to his office. It was a pleasant surprise when a few weeks later he invited Marion and me to his home and said that he would like to publish a French edition of *Principles of Heat Transfer*. This was now the fifth language into which my book had been translated, with previous editions in Spanish, Portuguese, Italian and Russian already in circulation. These translations made my book very accessible to students in many parts of the world and I was often surprised to find that students in foreign countries knew of me through my textbook.

During this period of my stay in France, the energy picture of the world had improved considerably and there was no longer a shortage of oil on the horizon. France had already begun to take an intense interest in nuclear power and was beginning to install nuclear power plants to satisfy a large portion of its energy needs. Moreover, France had relatively good relations with the Middle East and was able to obtain adequate liquid fuel for its transportation system. I spoke with French engineers about the future of obtaining energy from sustainable sources. They were professionally interested, but not to the point of teaching a solar energy course in any of the French universities.

THÈSES

présentées à la

Faculté des Sciences de l'Université de Paris

pour obtenir

Le Titre de Docteur de l'Université

par

Franck KREITH

1 ère THESE : Transfert de chaleur dans le cas d'un écoulement radial entre deux disques parallèles.

2 ème THESE : Propositions données par la Faculté.

———————

Soutenues le   1965 devant la Commission d'Examen

M. Ed. A. BRUN..........Président

MM. R. COMOLET.......⎤
                     ⎬ Examinateurs
S. F. SHEN........⎦

———————

PARIS 1965

My thesis defense was scheduled for the same day as that of another student at the University of Paris by the name of Henry Viviand. Two days before the thesis defense, Henry called Marion and suggested that we plan for some refreshments to be served to the examiners from the university. I am not sure that we knew exactly what was expected, but when we began to talk about the food that could be served, Mr. Viviand said "Never mind. This is not that important. The most important part of the refreshments is to decide whether to serve Champagne dry or medium dry." I was not accustomed to Champagne and left the choice to my French colleague who obviously picked "brut" (French for dry). The French do know how to live, and we both passed easily while the examiners were enjoying the Champagne.

After returning from my sabbatical leave in France, I proudly showed many people in the college the diploma of Doctor of the University of Paris, which I received for my research at the end of my sojourn. A lot of the people raised their eyebrows because a doctoral program in the United States takes several years of course work and some of the people could not understand how I was able to complete a doctoral program in such a short time. Among the skeptics was an investigative reporter from the *Denver Post* by the name of David Metzger. He had written an article about an employee in the Denver bureaucracy who had lied about his educational background and, as a result of this exposé, was fired. A few weeks after the semester started, I received a phone call from a person in the administration who said, "Frank, an investigative reporter by the name of Metzger is looking at all your personnel files and correspondence. He has a letter from the Dean of Engineering asking that whoever is responsible for the files should cooperate with the investigation." Dean Peters was aware of the investigation that Metzger had previously undertaken and which resulted in the dismissal of the person who had faked his degree. I realized that this move was just another of Dean Peters' efforts to harass me personally and force my resignation. There seemed no way for me to persuade the university administration to stop the investigative reporter from accessing my files. Consequently, I asked for help from one of the most prominent attorneys in Boulder. He decided to go directly to the owner of the *Denver Post* and warned him that since my Doctor's degree was official, any stories regarding a fake doctorate would result in legal action for slander against the newspaper. After some conversations between the lawyer and the owner of the newspaper, I was told that the newspaper had decided to abrogate the investigation and would not publish anything regarding my academic credentials. The matter came to the attention of the provost, Professor Lawson Crowe. He telephoned me the next day and told me "Frank, in my opinion it is pretty ridiculous to try and start this kind of investigation of a full professor with tenure who has served this university with honor for many years." I called Max Peters but he never admitted that he had any part in this nasty academic drama, which clearly was intended to embarrass me publicly.

CHAPTER 22—SOLAR ENERGY IN THE REAL WORLD, 1967–2000

Despite the disinterest in my work from the dean and the chairman of my department, I continued my research and professional publications. However, I decided to look outside the University for additional income because the dean refused to increase my salary. Fortunately, the heat transfer book was not only a professional milestone in my life, but it also made it possible for me to become financially independent of the vagaries of academic life at CU and the ever-changing politics of the bureaucracy in Washington. One of our friends, Ted Smith, was a realtor and he knew when my royalty checks came in; a day or two later, he would come to our house with a selection of two or three houses on the market that he believed could be purchased with the royalty income as down payment. He was persistent and always took us to houses that he felt were "good income property." These houses also gave me an opportunity to test my ideas about solar energy in the real world. Most of them were fairly old, but were structurally sound and could be made quite livable with minor modifications. I was very suspicious of the stock market, and Ted had little difficulty in convincing me to invest in Boulder real estate. At the time, houses could be bought with as little as a ten or twenty percent down payment, which was roughly the amount I received in a royalty check My income at the university was sufficient for the day-to-day needs of my family and Ted usually convinced me to use the royalty to buy one of the houses he showed to Marion and me. After we purchased a house and Ted received his commission, he took care of finding renters and using the rental income to make the monthly payments on the loan, as well as pay for electricity, gas, and water.

The first rental house we bought in 1967 was an old wooden structure, built in the late 19th century. It had been converted from a farm house into a four-unit apartment rental. Marion and I tried our hands at fixing up the place, which had been run down by previous tenants. We started by first steaming the old wallpaper and then peeling it off, hoping to paint over the surface. We had a lot of fun doing this job because under the walls we discovered old newspapers that had been used as thermal insulation. One of these had a headline proclaiming somewhat prematurely that John Dewey had been elected as the new U.S. President! (So much for trusting newspapers.) But we soon discovered that we did not have the skill to renovate such an old place alone and had to look for professional help. One of the people with whom we made contact, stayed with us as the upkeep contractor and is still helping maintain our house. Marion believes in keeping up relationships!

THIS IS THE SITE OF THE HISTORIC FLOWER HOUSE BUILT IN 1896 AND GIFTED BY FRANK AND MARION KREITH TO MENTAL HEALTH PARTNERS IN 1995.

A unique feature of that property was that an older miner from the days of the mining boom in Colorado lived in the house and had converted the entire yard into a flower garden. Marion and I decided to gift this unique structure to the local mental health organization, who placed the plaque on the building before it was renovated.

When we bought a property that had several apartments, Ted would, after a while, simply turn over the day-to-day management to one of the renters, but keep track of the overall management. This arrangement worked really well until Ted became ill with a heart condition and fairly unexpectedly passed away. From that time on, the management of the real estate became the responsibility of Marion and me. Both Marion and I were teaching at the University at that time. Marion was teaching French Literature in translation in the Honors Program and I was very busy with my engineering work. At some point, Marion made the decision to abandon her efforts to obtain a PhD in French Literature and took over full-time management of our real estate instead. I think she always regretted this decision a little, but found that she was a very good day-to-day manager of our real estate, while I continued to oversee major decisions such as updating the structures. Marion enjoyed decorating the interiors of some of the old apartments and she was very good at it. She also established friendly relations with our renters and some of them became our friends. We could, of course, not foresee the future development of Boulder; but as it turned out the value of the real estate appreciated considerably, gave us a steady income, and eventually made us fairly well to do in later life.

My first experiment with renewable energy systems in the real world was to install a retrofit solar domestic hot water system on our own house in the late 1960s. It was patterned after the systems I had seen in Israel, but required some modification to prevent the fluid passing through the collector from freezing in the winter. The collectors were bought from Novan, Inc, a local small business that sprung up after the

Carter administration began to offer tax credits for renewable energy systems. The installation was done by another small business specializing in domestic solar hot water. The system worked according to expectations and is still functioning 50 years later. The only interruption, other than normal maintenance, was a baseball from our neighbor's son that broke one of the glass covers on a collector on the roof. This successful solar venture gave me the confidence to try installing similar solar systems on some of the houses we owned or built.

My first major effort to install a solar hot water system on a commercial property was in the early 1970s. At that time, I owned a single family home with a shed on 9th street in the center of Boulder, one of the properties that Ted Smith had encouraged us to purchase. It was located in an area that permitted apartment houses to be built, and, in association with a builder and an architect, we designed a six-unit apartment house that we called Solar Six. As a part of my providing the land for this venture, I asked that the apartment house have a solar collector for domestic water heating on its roof. The building was an unqualified success, but in order to be competitive with other condos of the same size and quality, it was not possible to add the additional cost of the solar system to the sales price of the condos. Consequently, we had to absorb the cost of the solar system and there was very little profit on the sale of the condos. However, the solar system worked according to plan and operated successfully for many years.

Our involvement with buildings and solar energy was interrupted in 1975 when tragedy struck our family. My son Michael was enrolled in law school at Tulsa University when one day the counselor called us and said "Your son had a mental breakdown and I recommend that he should return home." We were totally unprepared, but naturally followed the counselor's advice. After Mike came back to Boulder, he was misdiagnosed as being schizophrenic and given some medication that almost killed him. Fortunately, when we took him to the CU Hospital, Dr. Friedman properly recognized that he was bipolar and helped him to recover. Mike has heroically

tried to lead a normal life with the help of antidepressant medication. He married an Irish woman, lived in Ireland for many years and became a basket maker in the country . He has a son, Daniel, who is studying law in Galway and we hope he will do what illness prevented Mike from achieving. Mike returned to Boulder in 2008 when his marriage fell apart. The photo shows all three Kreith men in a happy moment.

The second major solar installation was on a small development project that we designed in the 1980s with some friends—a builder by the name of Peter Brady and a local architect. I had purchased two old single-family buildings that were located in downtown Boulder on 2117 Walnut Street, an area that permitted the construction of multi-unit apartments. We had just returned from London where we had seen many small places called Mews. This gave us the idea to call our development Walnut Mews. Once again, I insisted that the development incorporate a solar domestic hot water system, which the architect and the builder agreed to do. When the building was completed, it received an Award of Excellence from the City of Boulder for exemplary contributions in architecture, conservation and general excellence of community design. But in 1981, the economy suffered a downturn and purchasers were only willing to pay for the size and quality of an apartment, but not for the cost of the additional solar hot water system that would save them money on the utility bill in the future. Thus, the venture was an architectural, but not an economic success.

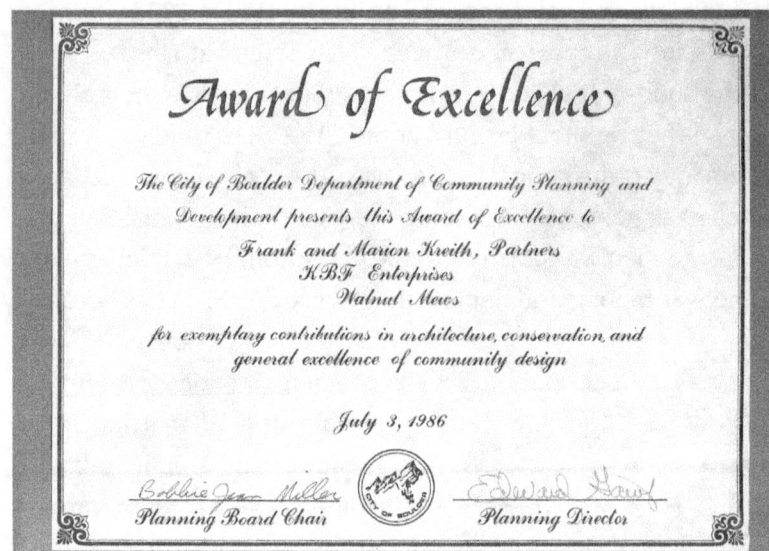

Award of Excellence

The City of Boulder Department of Community Planning and Development presents this Award of Excellence to
Frank and Marion Kreith, Partners
KBF Enterprises
Walnut Mews

for exemplary contributions in architecture, conservation, and general excellence of community design

July 3, 1986

Bobbie Jean Miller
Planning Board Chair

Edward Gawf
Planning Director

It became apparent to me that people buying houses or apartments only look at the initial sales price, but are not willing to pay extra for future reductions in their utility bills. In both cases, the future finances of the solar complexes were decided upon by a homeowners' association. As is normal in a solar hot water system, after a few years of operation the collector tank may begin to rust, the bearings on the water pump may begin to wear out, the antifreeze has to be replaced, and some additional minor maintenance work has to be conducted. All of these measures cost money. Although they would repay their cost in terms of savings on the utility bills within three to four years, the homeowners' associations in both apartment complexes refused to approve the expense of maintaining the solar systems. As a result, the systems ceased to function properly and were disconnected before they were able to repay the full initial investment by savings on the utility bills.

I believed that one of the problems with commercializing domestic hot water systems is that there is no simple instrument that shows the owner how much energy and money he or she is saving in the utility bills. In order to provide an indication of the monetary savings with solar heat, it is necessary to measure both the temperature rise achieved, as well as the flow rate of water passing through the solar collector system. These two items have to be multiplied and combined with the cost of electricity or natural gas necessary to provide the same amount of heat. Such an instrument would give a cumulative indication of the amount of money saved by the solar installation in real time and provide an incentive for the owner to pay for the upkeep of the system. This is one of the advantages of a photovoltaic solar system in which the electric power is both measured and recorded in real time so that one can easily determine both the energy as well as the monetary savings.

A third opportunity to invest in a solar thermal system occurred when one of my former students, Ken May, formed a small company called Industrial Solar Technology. The company built and installed solar thermal water heating systems in industrial complexes. One of these installations was on a gym to provide hot water for the showers of the inmates of a prison. In order to finance the initial cost of the system, Ken asked for investments from various people at $5,000 each. In exchange for this investment, Ken promised to repay investors in accordance with the amount and value of the heat delivered from the solar system. Since this was a major installation, Ken measured the flow rate and temperature rise of the solar-heated water separately and then calculated the monetary savings in the cost of natural gas which would have been necessary to provide the same amount of thermal energy as was produced by the solar collectors. Ken then collected the equivalent amount from the owners of the installations and distributed the proceeds in accordance with the investment agreement to construct the system. The cost of maintenance was subtracted from the available funds, which were distributed to each of the investors. This scheme proved to be exceedingly viable because it was properly maintained and the investment was repaid in eight years. After that time, of course, all of the savings from the solar system was pure profit to the investors. Ken repeated this type of investment strategy with a number of other installations and I believe that it provides a scheme for making solar systems viable because the installer provides a monthly or yearly accounting of the energy savings, as well as the financial savings, and mostly importantly, there is a direct interest in maintaining the system over the years in order to profit fully from its installation. Thus it removes some of the obstacles that were encountered in installing solar hot water systems in residential properties where buyers only consider the initial cost of the dwelling and a homeowners' association is looking only at the monthly bills rather than at the long term utility savings. All of these solar systems, particularly in commercial real estate, benefited from various incentives such as tax credits. But due to the vagaries of Washington politics, these incentives were periodically changed or discontinued, subsequently causing financial havoc for the many start-up companies who counted on the incentives for their bottom line.

I learned a lot from my experience of installing solar energy systems in the real world. First of all, only one of the small start-up solar companies that participated in the efforts to commercialize solar hot water systems survived after the Reagan Administration removed the tax credits. It was sad to see so much effort go to waste as a result of a lack of empathy for our future generation. But the technology did survive and is again growing today.

After several years of experience in real estate investment, I had gotten the hang of how one can use a small down payment to acquire real estate from which the income would pay for the interest and upkeep, as well as pay back the initial cash investment. Marion had a knack for making old buildings look attractive and using her sense of good interior decoration to make the houses and apartments look appealing. Her motto was, "If I couldn't live in the place, I don't want to rent it." With that approach, she had a gift of making friends with many of our renters and we are still in contact with some of them. However, at one point in our real estate acquisition process, we asked ourselves whether we really wanted to spend our time in this endeavor. I think that if we had continued our investment policies, we would be multi-millionaires today. But, it was our decision to do other things for the rest of our life. I decided to invest my time and effort in writing and teaching about solar energy rather than being a solar landlord. Fortunately, the real estate we had acquired gave us a sufficient income so we did not have to worry about vagaries that life could dish out financially.

CHAPTER 23—RESEARCH AT THE UNIVERSITY OF COLORADO, 1967–1974

Starting about 1965, I became interested in the energy and water budget of ecological systems, a topic that later proved to be useful in my work on energy sustainability. My interest in this topic started after meeting David Gates, a most unusual PhD physicist at the National Bureau of Standards in Boulder. His father had been a famous biologist and David continued that work by applying physics to ecology. He was interested in my work because he thought that heat and mass transfer were important factors in the equilibrium of ecological systems. Our first co-operation was in studying the micro environment of plants. My interest in this field increased when David thought about the possibility of regulating the water needs of ecosystems by chemical means. The potential for water conservation through transpiration reduction seemed like a topic of global importance because at least 99 percent of the water absorbed from the soil by plants is transpired into the atmosphere.

Water is one of the essential needs of all people. Under David's guidance I began to learn about the intricate nature of water transport in ecosystems and plants. It would be very difficult to appreciate the manner in which we viewed the potential of controlling water loss in ecosystems without some understanding of the basic plant structure and I ask for the indulgence of the reader to follow a simplified explanation of the transpiration process in order to understand what we were hoping to accomplish.

In 1972 Professor J. Anderson, on leave from the biology department at Idaho State University, joined our team in the laboratory at CU and we studied details about how the diffusion of water vapor, i.e., the water loss, can be controlled by chemicals called anti-transpirants (ATs). These ATs may be either metabolic and induce a closure of the stomata, which are the gateways of the plants water loss to the atmosphere, or be film-forming chemicals, which coat the leaf surface with a film that is impervious to water vapor diffusion. Metabolic ATs increase the resistance to carbon dioxide diffusion by photosynthesis as well as the water loss by transpiration. Thus, both CO_2 and water vapor transport are reduced, but carbon dioxide encounters an additional resistance in series between the leaf sub-stomatal cavity and the chloroplasts. If these two resistances are of similar magnitude, increasing the stomatal resistance causes a larger reduction in transpiration than in photosynthesis.

The transpiration process can be described by looking at a cross-sectional view of a typical leaf.

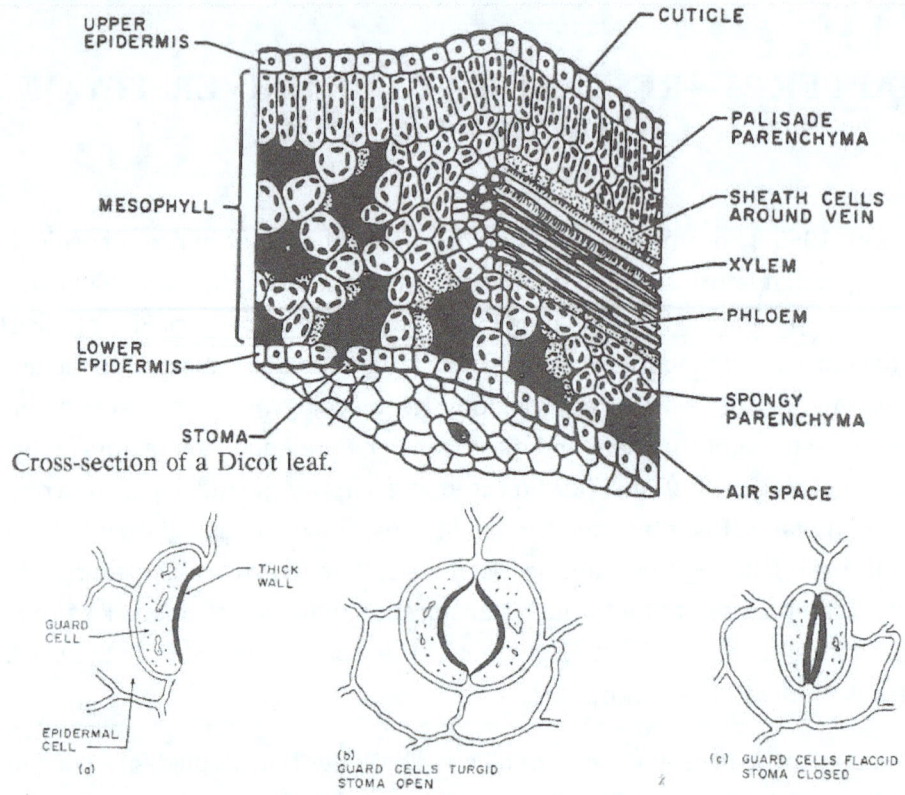

Cross-section of a Dicot leaf.

Since the outer walls of the epidermal cells are covered with a wet substance called cuton, which is almost impervious to the transportation of water, the epidermis and cuticular layer constitute a highly effective barrier to evaporation. Consequently, most evaporation occurs through tiny openings in the leaf called stomata. The size of a stoma is determined by the shape and relative position of two kidney-shaped guard cells surrounding it. A typical stoma is about 15 microns long, with a width up to 5 microns. Stomata occupy between 1 and 3 percent of the total leaf area in most green plants. Stomata openings increase when the guard cells become more turgid, swell, and bend outward, somewhat like a spring. The degree of opening of the pore is a function of the turgid pressure difference between the guard cells and the surrounding epidermal shelves.

My PhD student, Ashok Taori, and I developed a theoretical model for predicting the water loss and also studied a number of chemicals in their effectiveness to control the stomatal opening in an ecological wind tunnel at the University of Colorado. The PhD thesis by Ashock Taori provided quantitative estimates of the water reduction and related that to the stomatal opening by means of photographs.

While this research by Doctor Taori was in progress at the University of Colorado, I took a leave of absence to work in Israel on the fledgling solar industry there. However, I was also invited to give a progress report on our anti-transpirant work at the First International Congress of Chemical Engineering in Teheran, Iran. I described the process in a talk entitled "Chemicals Effecting Water Balance in Plants." It was published under the title *Effective Barriers: Chemical Anti-Transpirants on Heat and Mass Transfer in Environmental Systems* by Elsevier in 1974. There was a great deal of interest in the results of our experiments, but it appears that the only practical applications were for film-type anti-transpirants to reduce the evaporation from lakes and open water storage. However, the concept of reducing evapotranspiration without destroying the plant is again receiving interest, as water in the 21st century is becoming an increasingly scarce commodity.

At the time we were studying the water consumption of ecosystems, the U.S. government became aware of the political potential of using ATs on phreatophytes for water shed regulation. Phreatophites are spindly plants with deep roots that grow along watersheds. If the amount of water used by these plants is reduced, the water flow is increased without destroying the plants' root structure. This process could therefore be used as a water management tool without causing erosion resulting from an eradication of the roots of these plants. This idea was supported by a grant from the Bureau of Land Management in cooperation with the Office of Water Research and Technology in the U.S. Department of the Interior [OWRT grant B-099].The Bureau of Land Management was interested in applying the concept of reducing transpiration by means of chemicals along the Colorado River because the United States has a treaty with Mexico which limits the salinity of the water in the Colorado River entering Mexico. It was hoped that the use of anti-transpirants could reduce the amount of water taken up by phreatophytes along the river bed and thereby increase water flow and reduce the salinity downstream before the river crosses into Mexico. This would make it easier to meet the treaty obligation. A major effort in that direction was attempted by means of helicopters spraying the anti-transpirants. Unfortunately, it turned out that most of the stomatal openings of the phreatophytes are located on their lower surface and the spraying by means of helicopters did not prove to be effective. In order to effectively apply the anti-transpirants, it was necessary to have people walk along the river bed and spray the plants from below by hand. This proved to be difficult and uneconomical and thus the application of anti-transpirants for this application was not effective.

CHAPTER 24—NUCLEAR STIMULATION OF NATURAL GAS IN COLORADO, 1973–1975

"They shall beat their swords into plowshares…" –Isaiah 2:4

After my disastrous interaction with the Atomic Energy Commission at Princeton, I did not expect that I would ever have to face that behemoth again. However, life has strange twists and turns, and in the early 1970s our paths crossed once more. The reason for my second interaction with the Atomic Energy Commission (AEC) was the detonation of nuclear bombs in the Piceance Creek Basin of Colorado. At the time, Dr. Catherine B. Wrenn of the CU Political Science Department and I were co-directors of a research project on the interaction between technology, law, and politics funded by the National Science Foundation's program RANN (Research Applied to National Needs) at the University of Colorado. The goal of this project was to conduct interdisciplinary case studies of important emerging technologies that would have an impact on society. Major participants in this project included Professor Ronald E. West of the Chemical Engineering Department, Professor Donald Carmichael of the School of Law, and Professor William Winter of the Political Science Department.

Cathy Wrenn was a tall and athletic woman married to a professor of English. She was an outdoor woman and an excellent tennis player. We often played together in doubles' tournaments. Her outstanding tennis quality was a second service ace. I still have some of the trophies we won on my desk. Cathy was also an ardent environmentalist interested in projects that could help preserve the Colorado environment. When we applied for this grant, we were not aware that the Atomic Energy Commission planned to use atomic devices to stimulate natural gas production in the oil shale area of Colorado. However, this technology provided a significant and timely opportunity to study the interaction of an emerging technology with society in a real-world setting. The significance of such studies is today widely accepted and is emphasized in the preface to a book Dr. Catherine Wrenn and I later wrote. In that preface, U.S. Senator Floyd K. Haskell said "There is no mechanism for independent analysis of the true environmental, social, and economic costs of a new technology. The public has to rely on the opinion or bland assurances of those committed to a course of action." This condition certainly prevailed in the case of the Plowshare Program, where the big guns of the Atomic Energy Commission, including such famous people as Dr. Edward Teller who is known as the "father of the H-bomb," were frantically looking for opportunities to use nuclear energy for peaceful purposes. This effort was named Plowshare after the biblical reference.

The Plowshare Program was conceived in 1957. It was based on the premise that peaceful applications of nuclear energy could yield many benefits for human society. At that time, the potential dangers of radioactive contamination of the environment by nuclear applications was still being debated among experts. Some of them predicted that nuclear energy could be used for earth moving, mineral exploitation, energy production, and desalination of seawater. For us, the nuclear explosions in Colorado presented an ideal opportunity to look at the conflict between what some perceived as a vast nuclear potential and others saw as the possible adverse environmental impacts on citizens affected by nuclear energy.

The enabling legislation for the Plowshare Program was passed by the U.S. federal legislature in 1954 by an amendment to the Atomic Energy Act of 1946. It gave statutory existence to the desire that this newfound source of energy "be put to peaceful uses for the benefit of all mankind." On May 5, 1955, the AEC announced the start of two new nuclear programs. The first, the Power Reactor Demonstration Program, provided substantial federal assistance to utility companies to stimulate the building of nuclear reactors to create electric power. The second program, called Plowshare, envisioned the use of nuclear explosives in industrial applications. Initially the most promising application of nuclear devices identified under the Plowshare Program was in the broad area of mining, but was later directed specifically to the nuclear stimulation of natural gas in tight rock formations, which is called fracking today.

Early on, it became apparent to our research group that the AEC had a dual role in its legal authority: both to promote and to regulate nuclear matters on an operational level. In other words, the AEC was expected to sell the commercial use of nuclear devices, while at the same time it was charged with safeguarding the public against any adverse effects resulting from the use of these devices. Our legal expert regarded this as a potential conflict of interest, a position that became even more apparent as the Colorado nuclear project developed.

After having been barred from participating in the nuclear electric power program, I was initially quite intrigued by possibly being useful to the Plowshare Program, in particular when it became apparent that one of the first large scale tests to promote nuclear stimulation of natural gas was to take place in Colorado. The fracturing of tight formations by injecting a mixture of water and sand had been practiced for many decades without any objection from the public. The AEC thought that using nuclear devices for this purpose might be more effective. Nuclear stimulation of natural gas started with a project called Gas Buggy, which took place on December 10, 1967, in New Mexico. For this preliminary experiment, a 29-kiloton nuclear device was detonated 4,000 feet below the surface in the San Juan Basin by means of a fusion bomb. Although the AEC claimed that overall Gas Buggy had been successful in increasing natural gas production, it did admit that the gas had radioactive concentrations, especially of tritium, that according to the AEC "may necessitate the design of a new explosive device to minimize its production and potential adverse effects." The next industrial AEC venture was Project Rulison, the stimulation of natural gas in Colorado. It was designed to use a fission rather than a fusion bomb in order to reduce as much as possible to emission of tritium. The Rulison project was undertaken jointly by the AEC, Austral Oil Company, and the U.S. Department of the Interior, with CER-Geonuclear Corporation of Las Vegas acting in the role of program management. The Los Alamos Scientific Laboratory provided the technical direction. Project Rulison occurred on September 10, 1969 in Garfield County near Rifle, Colorado by the detonation of a 40-kiloton atomic bomb (euphemistically referred to as a device) at a depth of 8,430 feet. The bomb was more powerful than expected and scientists later determined that the blast was at the level of 44 kilotons, or the equivalent of 44,000 tons of TNT. By comparison, the bomb that destroyed Hiroshima on August 6, 1945, had a yield of about 15 kilotons. The bomb had been placed in a layer of solid rock, and once detonated, vaporized the rock all around it, carving out an empty cavity 350 feet high and 76 feet across. It fractured rock 400 feet in all directions, freeing natural gas as expected. However, the gas was also rendered radioactive and millions of cubic feet had to be burned off at the site because it was too dangerous to sell.

Even though the AEC claimed that nuclear fracking in Rulison had successfully demonstrated the potential of nuclear stimulation of natural gas technology, the AEC proposed a third nuclear stimulation blast called Project Rio Blanco. As in previous Plowshare projects, AEC combined with private companies, including the Equity Oil Company of Salt Lake City, Utah, and the now familiar CER Geonuclear Corporation of Las Vegas, Nevada. For the Rio Blanco experiments, three newly developed 30 kiloton Minata nuclear devices were to used in an effort to create three interconnecting chimneys one on top of the other, into which gas could flow and then be brought to the surface. The sponsors envisioned, assuming that the three-device test proved successful, detonating as many as 350 nuclear devices in the commercial phases of full development of the gas field. Although a number of political figures such as Governor Love, U.S. Senator Gordon Allot, and the newly elected U.S. Senator Floyd K. Haskell were against it, Project Rio Blanco was proceeding as planned. The only unexpected development was that oil shale companies in the area made known their opposition to nuclear stimulation of natural gas because they feared that this might handicap their subsequent exploration of oil shale to produce liquid fuels. Of those not involved in any economic way, Dr. Edward Martell, an atmospheric geochemist at the National Center for Atmospheric Research in Boulder, representing the Colorado Committee for Environmental Information, questioned the AEC conclusion that there would be no significant radiation damage to either subsurface water or local residents. During the month of May, a number of groups opposed to the tests. The Arapahoe Medical Society, following the lead of the 2,600 member Colorado Medical Society, resolved that the medical impact of the project required further evaluation and declared that potential risks exceeded the "reasonably anticipated gains."

We began our studies of the nuclear stimulation of natural gas as academic observers, but in the course of our four year effort, Dr. Wrenn and I found ourselves increasingly concerned about the potential hazards of large scale nuclear gas stimulation and critical of the manner in which the technology was presented to the public. In particular, as the controversy over the Rio Blanco project evolved in 1973, Dr. Wrenn and I departed from the role of uninvolved chroniclers and became part of the chronicle by presenting evidence at a hearing regarding potentially adverse impacts of using nuclear devices in the vicinity of populated areas.

Nuclear stimulation projects have been criticized for a number of reasons, but the chief concern is the fear of fallout radiation. Part of the concern stems undoubtedly from the average person's inability to understand the scientific complexity of the effects of radiation. In the last three days before the scheduled detonation, an appeal by an environmental organization of the previous Rulison suits was heard by the U.S. 10th Circuit Court. The district court decided that there were no grounds on which to base injunctive relief to prevent the nuclear detonation. In essence, the plaintiffs, namely the environmental group, had failed to present evidence to the court that any irreparable environmental damage would occur as a result of the nuclear experiment. A last ditch effort by a small band of protestors to prevent the detonation through acts of civil disobedience at the test site was taken care of by a police helicopter that removed the protestors from the site. In the end, neither the courts, the state of Colorado through its governor, nor civil protesters secured a postponement of the nuclear detonation

History thus seemed to repeat itself. In the final analysis, it came down to a judge facing conflicting testimony and opinions from the commercial and science community and basic disagreement among citizens as to the wisdom of pursuing the nuclear route to ease the national energy crisis. Without going into the detail, the presiding Judge Santos did not grant an injunction, and with the last legal roadblock to detonation cleared by that decision, the detonation occurred on Thursday, May 17, 1973. All of the residents within 7-1/2 miles of the test site had been evacuated and the people living within 7 and 14 miles of the site were requested to remain outside of and away from their homes or other buildings. Three nuclear devices were detonated simultaneously at depths of 5,840, 6,230, and 669 feet, producing an explosion with a force 4-1/2 times larger than the Hiroshima atomic bomb. Seismic shock waves from the blast registered a 5.5 magnitude on the Richter scale at the Colorado School of Mines in Golden. When the project sponsors proceeded with production testing, it was unexpectedly announced that another potentially harmful radioactive element, cesium-137, was contained in the natural gas and water emanating from the well. Hard on the heels of the discovery of cesium in the gas came an announcement from the State Department of Public Health that strontium-90, another radioactive material linked as a causative factor in cancer and leukemia, had also escaped from the nuclear chimneys. Moreover, the three underground cavities had failed to interconnect and, therefore, the original well was virtually useless for drawing off accumulated natural gas. In other words, the multiple blast technology had failed to live up to expectations. By 1974, the story was almost complete. The nuclear wells at Gas Buggy, Rulison, and now Rio Blanco had been sealed and there were no plans for their reactivation. The outcome of this major peaceful application of nuclear power was not successful and the process left me with a sense of unease about the social reliability of the nuclear establishment. A commission was established afterwards to summarize the nuclear plowshare project. This commission recommended that efforts should be made to use the conventional method of fracturing shale that had been used for decades to extract the natural gas from tight rock formations. But the financial success of this approach depended on the ability to drill horizontally underground, a technology that did not become available until many years later. One of the best known advocates of nuclear stimulation was Dr. Edward Teller. In the wake of the Arab oil embargo, the AEC saw an opportunity to apply nuclear devices as an instrument of mining domestic energy sources. Speaking to a Boulder audience in June of 1973, Dr. Teller said that although a large number of nuclear explosions would be necessary to stimulate natural gas, "fuel from Colorado could solve the nation's energy crisis." Dr. Teller's reference to the necessity of a hundred nuclear detonations in the state was frightening and many people in the state of Colorado began to take notice. In the wake of two underground nuclear experiments in their state, the voters of Colorado took matters into their own hands. They went to the polls in November of 1974 and passed a referendum that made it unequivocally clear that they would permit no further nuclear stimulation of natural gas without their prior consent. Thus, the people of the state accomplished what neither the federal government, nor the courts, nor the state and its governor were willing or able to do—to halt further underground nuclear blasting in the Colorado.

Frank Kreith
Catherine B. Wrenn

University of Colorado

THE NUCLEAR IMPACT

A Case Study of the Plowshare Program
to Produce Gas by Underground
Nuclear Stimulation in the
Rocky Mountains

WESTVIEW SPECIAL STUDIES ON TECHNOLOGY,
NATURAL RESOURCES AND THE ENVIRONMENT

The story of the AEC's detonation of nuclear bombs in Colorado presented here is quite cursory and incomplete. However, Dr. Catherine Wrenn and I published a 248 page book under the Westview Special Study on Technology, Natural Resources, and the Environment series. The book, entitled *The Nuclear Impact*, was reviewed by Dr. Morgan Sparks, the President of Sandia National Laboratories, who wrote "the core of the book is a well-balanced, thoroughly documented account of the events and controversies surrounding the three Plowshare experiments in stimulating natural gas production with nuclear explosives." We did not succeed in preventing these nuclear experiments, but hopefully the history documented in our book will in the future make it more difficult for a Dr. Strangelove to expose people and the environment to nuclear radiation or other kinds of potential dangers without prior consent of the public.

On a more personal basis, the involvement in the nuclear stimulation project had an enjoyable ending. I was asked by the Sierra Club to participate in a roundtable discussion about energy and the environment with Dr. Edward Teller, Amory Lovins,

the co-director of the Aspen Institute, U.S. Senator Floyd Haskell, and the well-known singer and environmentalist John Denver. Each of us was given five minutes to make a presentation. Dr. Teller spoke in favor of continued nuclear stimulation experiments, explaining that it would take more than a hundred nuclear devices to get the natural gas flowing. I spoke in favor of wind power, and Amory Lovins proposed more energy conservation. Senator Haskell stressed the importance of environmental impact statements. Finally, it was John Denver's turn. John had brought his guitar, and instead of a carefully reasoned analysis of environmental impacts of a technology, he simply began to sing one of his favorite songs about preserving the environment, "Rocky Mountain High." From that point on any further serious discussion ceased and the entire crowd simply wanted to hear more songs; John obligingly filled the auditorium with his music.

John Denver died unexpectedly a few years later when his small private airplane crashed. After his death, people came to see his ranch in Colorado and found that John not only had an airplane, but also two sports cars, one of which was a Ferrari, and an underground storage tank for gasoline, which was, as John Denver put it, " just in case Colorado should run short." John Denver was an ardent defender of the environment, but apparently his interest in assuring continued fuel for his airplane and sports cars took precedence in his own life. However, his music will continue to be an inspiration for us, and "Rocky Mountain High" is still one of my favorite songs.

CHAPTER 25—ENVIRONMENTAL CONSULTING SERVICES, INC., 1974–1975

Catherine Wrenn's and my efforts to study the effects of a new technology that some experts claimed to be economically lucrative and would lead to new energy sources, but at the same time could have adverse impacts on the health of people and the environment, led to recommendations that have had a lasting effect way beyond our Plowshare study. The core of our recommendations for the future of such projects centered on a process called a technology assessment (TA). The basic goal of a TA should be to improve and make more rational the means by which key public policy decisions are made in a democratic society. The assessment process is particularly important when large-scale government actions are taken which allocate public resources to control the development of a new technology, particularly in the energy field.

The goal and end result of a technology assessment should be to encourage developments which are socially, economically, and environmentally in the public interest. For projects in which detrimental side effects seem likely, the assessment process should investigate different options of achieving the project goals. If, after careful analysis, a project is deemed to be excessively dangerous and cannot be justified in terms of public interest and welfare, this information should be passed on to the appropriate decision makers for action. We further recommended that the technical assessment process should not only be a routine part of decision making, but should also give an opportunity for external checks on the government by independent experts acting as concerned citizens. Although technology assessment procedures are still new and imperfect, I believe that they can help to steer a course for making better public policy decisions in a democratic society. Our basic ideas have subsequently been proven correct, especially the recommendation to search for alternative options. After nuclear stimulation of natural gas was discontinued, better and more efficient methods for extracting natural gas by more conventional methods were developed. Moreover, the price of natural gas continued to escalate. As a result, private exploration companies returned to Colorado at the turn of the 20th century and began to extract the gas profitably without nuclear stimulation. On October 16, 2008, the *Boulder Daily Camera* reported that, according to current estimates, the Roan Plateau has 9 trillion cubic feet of recoverable natural gas and plans call for 1,570 wells drilled from 193 pads over 20 years, while preserving 51% of land on the Roan. That estimate of recoverable gas matched the expectations of the AEC, but the technology existed now to recover the gas without nuclear stimulation.

After the positive response to our recommendation to file environmental impact statements before any major energy project got underway, I decided that I would like to try my hand at an industrial or commercial application of this idea. Although I had never done this before, I decided to form a corporation specifically designed to prepare environmental analyses and assessments of major energy projects. The company was named Environmental Consulting Services. I informed the dean of the engineering college of this step and told him that I would use only the one day per week that a faculty member is permitted to undertake consulting services for work in this new corporation. The other members of the corporation included Dr. Jan F. Kreider; Professor Ronald West of the Chemical Engineering Department; Colorado State Representative Charles (Chuck) Howe; and an old friend and tennis partner by the name of Ed Harris, who had been a program manager in the aerospace industry. The most active of the group was my friend and former student Jan. After graduation from Case Institute of Technology, Jan took a job with General Motors as a test driver of the new car models' air conditioning systems. One of his favorite places to test these systems was the long and tortuous drive over Red Man's Pass in Colorado. He did not want to return to Detroit after tasting Colorado's fresh air and decided to enroll in the PhD program at the University of Colorado, although he was considerably older than the average student. Jan was not only an outstanding student, but a good technician with an inventive mind. I was happy that he chose me for his mentor and PhD advisor. He was also an ardent hiker and mountain climber and we made many memorable camping trips together. The most exiting of them was a three-week long trip down the Colorado River in dories, small four-man wooden boats similar to those used by Major John Wesley Powell in his famous exploration of the river in 1869. The rapids in the river are graded from 1 to 10, with 10 being the most powerful. Only one of the rapids is of grade 10, and Jan was the only one to go over it in the boat with the guide. The rest of us walked around it.

As a part of his investigation into the legitimacy of my doctorate, the *Denver Post* reporter, David Metzger, had found a letter from me requesting permission to found ECS and become its president. He showed the letter to Dean Max Peters and when the dean learned that I was to be the president of the corporation, a situation which in recent years had been widely repeated and accepted by the university, he called me into his office and threatened to dismiss me from the university because he claimed it was illegal for a faculty member to be an officer in a company, even if no more time than the allowed one day per week consulting services were spent on the work. I was unaware that the dean acted without legal authority, but was scared enough to ask the people in our new company for advice. It was decided that the best thing to do would be for me to step down as the president and appoint Jan Kreider to take my place. This seemed a reasonable thing to do, in particular since Jan had always been a rather aggressive, well-organized, and, of course, technically highly respected person. What I did not realize at the time was that both Jan and I had immensely strong egos that might clash. This should have been apparent to me from some previous problems we had as co-editors of the *Handbook of Solar Energy*. When Jan felt that his contributions to the book were more important than mine, he had asked the publisher to remove my name as the co-editor. It was only through the interaction of our mutual friend, Bill Begell, who had successfully published and promoted our previous three books, that we settled on a compromise whereby Jan was named Editor-in-Chief and I remained the co-editor.

Initially, we had started with a great deal of enthusiasm to promote the new corporation. It was our hope that in addition to filing environmental impact statements, we could also make people aware that renewable technology had a smaller environmental impact compared to fossil fuel energy generation. I do not remember all of the various companies we contacted and what our efforts to promote the company were; however, I do remember an event that today I feel is one of the saddest and most unfortunate events of my entire career. One day, I discovered that Jan had made some contacts and presentations to several outside sources without keeping me in the loop. Since it was my feeling that Jan's position as president had been forced upon me by the dean, I felt that he owed me the courtesy of keeping me informed. I was, therefore, angry and disappointed and called a meeting of the participants in the company. To my everlasting regret, I had let my ego run away with me and, instead of simply appreciating Jan's efforts, I accused him of having undercut me. This created some bitter verbal interactions among the various members of the company, and at one time Marion, who was also present, called him offhandedly an SOB. That brought forth a terrible reaction from Jan who screamed, "Whatever may have happened, one should not accuse his mother of being a dog." He walked out and I cannot blame him for his anger, even if the insult was not intended. But the end result of this row was that Representative Howe asked to be removed from the board and, after a few more days, it became obvious that due to the lack of confidence in Jan's leadership, the company had to be dissolved. This experience also made me realize that I was really not cut out to be the administrative head of a commercial organization and should restrict my efforts to what I was good at, which was writing, research, policy analysis, and teaching. I have tried to follow this realization for the rest of my life. It was a sad, but probably also instructive experience and I only wish that it had not led to the alienation between my one-time wonderful friend, former student, and colleague, Jan Kreider, and me. Jan and I continued our professional cooperation for some time and both Marion and I have made efforts over the years to heal this wound, but have never really been successful. Jan formed his own solar consulting company but over the past few years has totally withdrawn from all social contacts.

CHAPTER 26—PRESIDENT NIXON'S PROJECT INDEPENDENCE, 1973-1976

In the period between 1959 and 1965, I had started to take an interest in solar energy conversion, although the technical application of solar power at the time was directed largely toward the space program. I developed a graduate course on radiation heat transfer and design of solar energy conversion systems and constructed a single axis tracking concentrating solar collector on the roof of the engineering building at the University of Colorado. This collector, alongside an advanced version used in the construction of a solar power plant in Arizona, is shown below.

Then, I presented a short course on solar radiation for NASA engineers at the University of Oklahoma and used the notes for that course to prepare a textbook entitled *Radiation Heat Transfer for the Design of Spacecraft and Solar Power Plants,* published by the International Textbook Company in 1962. The book was later translated into Spanish and used as a text in many universities. Although the funding for these activities came largely from NASA's space program, it turned out that the work I did in this period was directly applicable to my growing interest in deploying solar technology for practical applications on Earth.

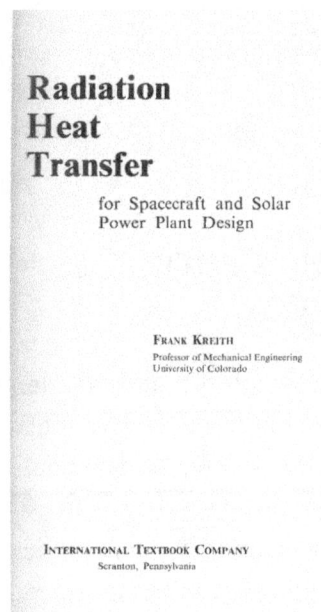

The modern era in federal energy policy began October 17, 1973, when the Organization of Petroleum Exporting Countries (OPEC) announced an embargo of their oil exports to countries supporting Israel in the Yom Kippur War. Congress, in three successive administrations, responded in the next five years with extensive laws and regulations in the expectation that assistance from the federal government would solve the energy problem, especially the U.S. dependence on oil imported from hostile countries in the Middle East. President Nixon was spurred into action in January of 1974 when oil prices suddenly rose from $3 to $14 per barrel. The United States, in which oil consumption had been growing while domestic reserves were dwindling, had become reliant on imported oil from the Middle East and was forced to negotiate an end of the embargo. The Nixon administration began negotiations with Arab oil producers to end the embargo and, in March of 1974, Arab oil producers decided to lift the embargo. In the meantime, Nixon had created the Federal Energy Office and appointed an "Energy Czar" with the power to allocate oil supplies. He also requested the preparation of a plan known as "Project Independence," which was to make the U.S. independent of imported oil by 1985. Previously, Nixon had stated in his energy message on June 4, 1971, "The sun offers an almost unlimited supply of energy if we can learn to use it economically." During the 1970s, the federal solar program grew and included not only R&D, but also joint participation with the private sector in demonstration projects, commercialization, and information dissemination.

My first major national involvement in promoting conservation and solar energy technology began in 1975 when Vice-President Nelson Rockefeller invited me to participate in a preliminary assessment of the national energy situation. My best recollection is that energy conservation was one of the key topics.

THE VICE PRESIDENT
WASHINGTON

October 23, 1975

Dear Dr. Kreith:

     Thank you for your splendid participation in the recent Denver Public Forum. Under Secretary Frizzell and Assistant Administrator Rosenberg have informed me of your outstanding testimony in the afternoon session on Resource Development and the Environment.

     I congratulate you on your eloquent contribution to the Forum and to our review of domestic policy. It was a privilege to have your insights and perspectives as part of the Forum. Such an honest and open exchange of views is a necessary part of our democracy.

     My warmest thanks to you and all who worked with you in preparation for the Forum.

     With best wishes,

     Sincerely,

Dr. Frank Kreith
1485 Sierra Drive
Boulder, Colorado 80302

In response to President Nixon's challenge in 1973 to achieve U.S. energy independence, the National Academy convened a panel of experts to study various options to meet this goal. The panel, entitled the Committee on Nuclear and Alternative Energy Systems (CONAES), was formed on April 1, 1975 by Dr. Robert C. Seamans, then-Administrator of the Energy Research and Development Administration (ERDA). A final report by this committee was transmitted to the then-Secretary of Energy, Charles W. Duncan, Jr., in December of 1979. The report stated that "the most critical near-term problem in energy supply for this country is fluid fuels." CONAES noted that "world supplies of petroleum will become severely strained in the 1980s and that serious problems are likely to occur as world oil production begins to peak." The committee further noted, "There are several plausible options for an indefinitely sustainable energy supply, potentially accessible to all of the people of the world."

I believe that a shortcoming of the report is its lack of emphasis on the effects of burning fossil fuel on the increase of CO_2 in the atmosphere causing global warming. The report states "if the worldwide combustion of fossil fuels, particularly coal, continues to increase, the problem [greenhouse effect] could begin to be perceptible as early as the first few decades of the 21st century, or it might not become significant until the latter part of the 21st century." The report also speculated that the effects of global warming might be postponed for a century. The lack of concern about global warming may, in my opinion, have influenced some of the conclusions and recommendations in the report.

The committee, however, was fully cognizant that moderating the growth of energy demand was essential. The report states that, "slowing the growth of energy demand will be essential regardless of the supply options developed during the coming decades. In fact, the demand element of the nation's energy strategy should be accorded the highest priority." The report also identified the near-term options for solar energy in agreement with my previous testimony at the White House Public Forum and the conclusions Jan Kreider and I presented in our book, *Solar Heating and Cooling*, in 1975. The book emphasized that solar energy could be used in the near-term for domestic hot water heating, providing industrial and agricultural process heat and low pressure steam, but also mentioned the potential of photovoltaic technology for generating electricity in the future.

Regarding the potential of nuclear power, the CONAES report states "nuclear power could make a substantial contribution to the base load electrical system of the United States in the intermediate term...Nevertheless, the expansion and further development of nuclear power face uncertainties and controversies." The report is generally supportive of nuclear power, but in my opinion, underestimated the increase in the cost of building nuclear power plants in the 21st century. Furthermore, it did not fully address the dangers of nuclear waste and the need to have a permanent depository for high-level nuclear waste before nuclear power can be expanded substantially. This has become a particularly disturbing problem because no safe nuclear depository exists at this time in 2014. It is also worthy of note that the committee did not include hydrogen energy among the options for energy independence. In fact, the word hydrogen does not even appear in the index. One can only conclude that the committee fully understood that hydrogen is not an energy source and should therefore not be included. It is thus very strange that another committee of the same Academy 30 years later recommended to then-President George W. Bush that a hydrogen economy was technically feasible. In summary, the CONAES report recommended in 1975 that energy independence could be achieved by means of conservation, synthetic fuels, effective use of coal and nuclear power to produce electricity, and the use of solar energy for low temperature heat. These developments and the promise of President Nixon to make the United States energy independent by 1985 created a sudden upswing of interest in developing indigenous renewable resources, particularly solar energy. As a result my solar energy activities in 1974 and 1975 became quite frantic. In the first six months, I presented eight lectures at places like NCAR, Argonne National Laboratory, Denver's Channel 9 TV, and the Shriners' Club. I also participated in a panel discussion entitled "Alternative Sources of Energy" at the Edward Teller Center for Science, Technology, and Political Thought in Denver. Then I joined forces with Dr. Jan F. Kreider and Dr. George Loef, an experienced solar engineer, to develop short courses on solar energy utilization. In the period between 1973 and 1974 we presented several solar energy short courses. Jan and I had been thinking seriously about how to promote the use of solar energy and were pleased to join with George Loef, who was at the time a professor of chemical engineering at Colorado State University. George had, in the early 1950s, personally built and operated solar heating systems on his own house at 1719 Mariposa Avenue in Boulder, 2 blocks from our first house in Boulder at 1911 Mariposa. He had to dismantle the solar installation after leaving the CU faculty because realtors found that they could not sell the house with its unconventional glass panels on the roof. However, George later built a second solar home in Denver, and his experience was quite influential in the way the short courses were structured. We presented our solar

photo courtesy of the Camera

Dr. George Lof's solar home at 1719 Mariposa Avenue in the 1950s

short course to audiences of engineers, architects, and city planners all over the country. Sometimes we were able to combine work with pleasure, as for example when we scheduled our courses during the winter in Aspen. This gave us, as well as our students, the opportunity of studying solar energy in the morning, and then buying half-day tickets for skiing on the slopes in

the afternoon. It is my recollection that overall we gave this course 22 times in two years.

At the time we developed the course, we believed that solar heating of buildings would be a good application for solar energy that would substantially reduce the need for heating oil and/or natural gas. However, some of the more practical engineers among our students pointed out that the use of solar energy for heating homes would not be as economical as applications for which the solar installation could function all year long, because the heating of homes was seasonal and the solar systems stood idle for the summer months. This practical input from our students was a very valuable insight that guided my decisions in structuring the thermal research program at the Solar Energy Research Institute some years later.

In the spring of 1974, the American Solar Energy Society (ASES) held an energy conference in Denver, where I met Professor Gershon Grossman from the Technion in Israel, an outstanding engineer and solar enthusiast who later also become one of my close friends. Gershon had come to Denver because Israel, which has abundant sunshine, was interested in developing new ideas in solar energy. I mentioned to Gershon that I was working at the time with one of my colleagues on a new concept of solar energy conservation, which was able to concentrate solar energy without the need of the reflecting surface to track the sun. The idea was called SRTA (spherical receiver with tracking absorber), and was initially tried out on the heating system for a house near Boulder where part of the roof was formed into a spherical shape covered with reflecting material and with a tracking absorber that was able to follow the rays reflected from the surface. This design was capable of generating sufficiently high temperatures in the cylindrical absorber to operate a small engine and, if installed as a power generator, could operate all year long. Gershon and I decided to explore the possibility of cooperating in the development of a commercial model and made a proposal to the U.S–Israel Bi-National Science Foundation, which encouraged cooperation between Israeli and U.S. scientists, for a grant. Our application was granted in 1976, and several SRTAs were subsequently built in Israel. The first had a 2.5 m diameter and a subsequent larger version had a 10 m diameter, as shown in the photograph. Despite its attractive feature of not requiring a tracking reflector, the idea turned out to be commercially unattractive because of the high maintenance costs of the prototype and the inferiority of spherical optics compared to other optical options such as the central receiver power tower.

In 1973, Bill Begell, the editor of a small company entitled Hemisphere Publishing, encouraged Jan and I to publish an edited version of the elaborate notes we had prepared for our short courses as a hardcover textbook. This was a great idea and we were able to complete the preparation of the manuscript within a short time. The edited version of our lecture notes was published jointly by Hemisphere and McGraw-Hill Book Company as a book entitled *Solar Heating and Cooling* in 1975. The book was a first in its field and was highly successful, being reprinted several times in the following years. It was also used as a text in many universities and translated into Japanese in 1976. I have a copy of the Japanese translation and the only thing I can read is Jan's and my names on the title page. The English version of the book was the featured selection of the Mechanical Engineers' Book Club in 1975, and the main selection of the Library of Urban Affairs in 1976.

DR. JAN F. KREIDER is President of Environmental Consulting Services, Inc., a Colorado-based engineering company specializing in energy development and environmental analysis.

He has several years of industrial experience in the design and manufacture of heating and air-conditioning systems with General Motors and is an expert on mathematical modeling of engineering systems. He has prepared numerical models for such systems as solar energy collectors, solar-heated homes, and air-dispersion patterns of the Colorado River Basin. Dr. Kreider has also published articles on engineering instrumentation and is currently a consultant to the National Center for Atmospheric Research and the Bureau of Economic Research at the University of Colorado.

He received his B.S. from Case Institute of Technology and his M.S. and Ph.D. from the University of Colorado.

DR. FRANK KREITH is a professor in the Department of Chemical Engineering at the University of Colorado at Boulder, where he teaches a course in the practical utilization of solar energy. He is also affiliated with Environmental Consulting Services, Inc.

The author of more than 100 published works in the field of heat transfer, including *Principles of Heat Transfer* and *Radiation Heat Transfer*, Dr. Kreith has been a guest lecturer at more than 45 universities throughout the United States and abroad. He has served on the advisory panel for the National Science Foundation on Solar Energy Utilization, and has been a consultant to the Prime Minister's Office of Israel, the Pollution Control Commission in Great Britain, and NATO.

Dr. Kreith received his B.S. and M.S. from the University of California, Berkeley, and his doctorate from the University of Paris. He has been the recipient of many awards, including a Guggenheim Fellowship, various Fulbright travel and lecture grants, and the 1972 Heat Transfer Memorial Award of ASME.

McGRAW-HILL BOOK COMPANY
Serving the Need for Knowledge
1221 Avenue of the Americas
New York, N.Y. 10020

0-07-035473-1

Partly as a result of the book and the wide range of students from different companies who attended our course, I received invitations from many parts of the world to assist as a consultant and advisor on solar energy projects. There is not space to relate all of the wonderful experiences, but I will come back to some of them. However, interest in developing solar energy was intense beginning in 1975, and I provided assistance to numerous solar projects in Canada, Japan, Taiwan, Turkey, Israel, Morocco, Yugoslavia, Sweden, Thailand, and India. In October, 1975, I was invited to present testimony at the White House public forum on domestic policy on the potential for solar and geothermal energy, and in December of that year, I was appointed Advisor to the NSF/ERDA Solar Energy Program and became a member of the NSF Solar Energy Proposal Evaluation Team for Advanced Energy Research and Technology in Washington, D.C. (1975). Unfortunately, when the price of oil dropped in 1975 and the energy crisis abated, the worldwide interest in developing solar and other renewable sources to replace fossil fuels diminished. Many of the universities who had embraced the idea of renewable energy and offered courses in solar energy conversion in the 1970s discontinued these courses because their graduates were unable to find jobs in the solar energy field. Also many of the small solar startup companies went bankrupt.

In the summer of 1976, I received a NATO Senior Fellowship which offered me an opportunity to take my enthusiasm for solar energy to Europe. As part of the fellowship, I represented the United States at the international UNESCO Solar Energy Conference in Geneva August 29–September 4, 1976, where I presented a paper on solar energy for developing countries. In the following months, I gave lectures on solar energy at the Technische Hochschule in Delft, Netherlands, the Imperial College of London, the University of Karlsruhe and the University of Stuttgart in West Germany, the Technische Hochschule of Munchen, and finally, at the Laboratorie Dynamics de Fluide in Portierre, France. These lectures gave me the opportunity to revive my knowledge of German and French and present the lectures in the language of the country. Upon returning to the United States, I gave another short course on Solar Heating and Cooling sponsored by the Solar Energy Division of ASME in New York and then resumed my teaching at the University of Colorado.

CHAPTER 27—PRESIDENT FORD, 1974–1976

Unfortunately, my enthusiasm for solar energy was not shared by the American energy establishment in Washington, which considered renewable energy technology too expensive for commercialization. Moreover, the political events in 1974 which forced President Nixon to abdicate after the Watergate break-in took center stage and diverted interest from energy. President Ford, who took the helm of the U.S. administration after Nixon abdicated, did not consider the goal of energy independence realistic and did not encourage solar energy development. But he realized that the inevitable problems with U.S. limited petroleum resources would need to be addressed. The Energy Policy and Conservation Act of 1975, the last piece of legislation signed by President Ford, established the Corporate Average Fuel Economy (CAFE) program which set, for the first time, energy-efficiency standards for cars and trucks. The requirement that the mileage of cars had to increase was the most successful federal energy conservation mandate for the transportation sector in the U.S. The mileage for passenger cars rose from 13.5 mpg in automobile model year 1978 to 27.4 mpg in 1985 and for trucks from 11.6 mpg to 19.5 mpg in 1985. The three-year average for cars and trucks is shown below. In the mid-1989s, the car makers Ford and GM lobbied the Reagan administration to lower the standards, which

History of Fuel Economy: One Decade of Innovation, Two Decades of Inaction

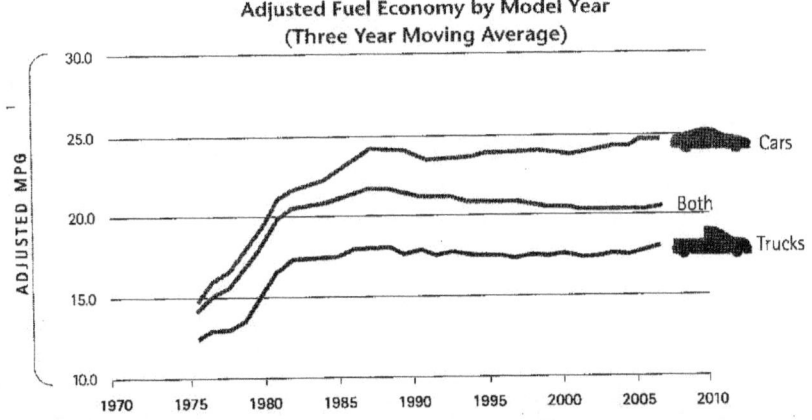

Adjusted Fuel Economy by Model Year
(Three Year Moving Average)

were dropped to 26 mpg in1988. During his term in office, President Ford also set up an emergency petroleum reservoir that is still in place.

At the behest of Secretary of State Henry Kissinger, President Ford initiated several efforts aimed at fostering international cooperation among energy consumers, including the creation of the International Energy Agency (IEA). One of its goals was to promote alternative energy sources. The enormous political resistance to taxing any kind of energy became apparent during this period, when John Sawhill, the head of the Federal Energy Administration, was forced to resign following backlash for his support for a five cent per gallon tax on gasoline.

One interesting event in my many solar energy activities at that time was to be a design consultant to the Bureau of Indian Affairs for a solar retrofit system of a gymnasium in the Breadsprings Middle School of the Navajo Reservation in New Mexico. For this project, I cooperated with my friend, Roland Winston, in the design of what at that time was an advanced compound parabolic concentrator (CPC) collector. The idea of a CPC collector was the result of a basic invention by Dr. Winston based on the geometric concentrator theory that defined the thermodynamic limit of concentration. Roland has, in the meantime, obtained a patent and made this concept a cornerstone of international development in solar energy. We have been told that the Navajo solar system we helped to install on the roof of the gym survived for many years until it was destroyed by stone-throwing vandals that broke the glass of the collectors.

CHAPTER 28—ESTABLISHMENT OF THE SOLAR ENERGY RESEARCH INSTITUTE, 1977–1979

The Solar Energy Research Institute (SERI) was originally conceived in 1974 as the United States emerged from the worst energy shortage in its peacetime history. The shortage of gasoline caused long lines at the pumps and resulted in the realization that the U.S. was nowhere near achieving energy independence as promised by President Nixon, but was becoming increasingly dependent on oil from the Middle East.

Congress, concerned then as now, about future energy supplies and overdependence on foreign petroleum sources, passed the Solar Energy Research, Development, and Demonstration Act calling for the creation of SERI. Later in 1974, the Energy Research and Development Administration (ERDA) began its operations. The Administration commissioned a study program to determine SERI's role, mission, and management methods. Based on the Congressional mandate and on results of the studies, ERDA decided to establish this new solar energy research institute to provide major national focus and coordination for solar energy efforts.

When I heard about DOE's requests for proposals for a solar energy research institute, I made an effort to bring the institute to the University of Colorado at Boulder. I thought this made sense because Colorado has an excellent solar climate and Boulder is the location for the National Center for Atmospheric Research and the National Institute for Science and Technology, as well as many advanced startup companies in computers and renewable energy. I searched for similarly minded people among my colleges and found support from both Governor Richard Lamm and U.S. Senator Gary Hart. We jointly submitted a letter to the U.S. Energy and Development Administration suggesting that the University of Colorado administer and house the future Solar Energy Research Institute.

Friday, May 9, 1975   **DAILY CAMERA 9**

Solar Institute Sought For State

WASHINGTON (AP) — Colorado's congressional delegation and Gov. Richard Lamm have submitted a concept and recommendation report to the U.S. Energy Research and Development Administration asking that a national solar research institute be established in Colorado.

Last year, Congress passed the Solar Energy Research, Development and Demonstration Act which directed creation of the institute.

In a letter accompanying the report, the group said it is unanimous in the desire to have the institute in Colorado and efforts are being coordinated to achieve that goal.

"In Colorado, we are particularly sensitive to the extreme impact of conventional sources of energy and to the increasing problem of depletion of our nonrenewable resources," the letter said.

Unfortunately, when Max Peters, the dean of the college of engineering heard about our efforts, he nixed the idea because he thought that research to develop oil shale and nuclear power were much more promising than solar energy. It was a disappointing decision for many of us, and I still believe that with administrative support it would have been possible to find a home for SERI within the Colorado university structure.

The studies commissioned by ERDA recommended that instead of being established as a federal laboratory, SERI should be operated by a private entity under contract to the federal government. Accordingly, in March 1976, ERDA issued a request for proposals calling for a management plan, a management team, and an initial site with an option for a future site. In July 1976, ERDA received 20 proposals offering sites in 16 states. In March 1977, after extensive evaluation, ERDA awarded Midwest Research Institute the contract to establish and operate SERI at a site in Golden, Colorado. The Solar Energy Research Institute formally opened on July 5, 1977. In October, 1977, ERDA's functions were assumed by the newly-created Department of Energy (DOE), which undertook the future supervision and funding of SERI.

Planning for the Thermal Conversion Branch of the Solar Energy Research Institute began in November of 1976, when Mr. Crozier, the Personnel Director of MRI, called me in London where I was spending a CU faculty fellowship to complete my new book entitled *Principles of Solar Energy Engineering*. He asked if MRI could place my name on the MRI proposal and if I would accept the responsibility of directing the Research and Development of Solar Thermal Conversion if the bid of MRI to manage the institute was successful. Needless to say, I enthusiastically agreed.

Frank Kreith / Jan Kreider

A breakthrough text on the engineering design of solar heating and cooling systems from McGraw-Hill/Hemisphere.

The first totally comprehensive text on the design and analysis of successful solar energy conversion systems.

I completed the new solar energy textbook with Jan Kreider after returning to the U.S. in 1977. It has subsequently been widely used all over the world as a text in solar energy courses. The cover for the book is a copy of a painting by the Yugoslav primitive painter named Ragusin which Marion and I liked so much that some years later we set out on a long trek to find the original.

My enthusiasm for participating in a proposal to establish a Solar Energy Research Institute was buoyed further when President Jimmy Carter outlined his proposed energy policy in a televised speech on April 18, 1977. He began with the historic statement, "With the exception of preventing war, this [the energy crisis] is the greatest challenge our country will face during our lifetimes. The energy crisis has not yet overwhelmed us, but it will if we do not act quickly." He went on to say that, "We simply must balance our demand for energy with our rapidly shrinking resources…This difficult effort will be the moral equivalent of war." He continued by reminding the country that, "Although the 1973 gasoline lines are gone and our homes are warm again, our energy problem is worse tonight than it was in 1973." He then warned that,

"Unless profound changes are made to lower oil consumption, we now believe that early in the 1980s, the world will be demanding more oil than it can produce." With foresight that today only a few people can recall, he stated that, "If we fail to act soon, we will face an economic, social, and political crisis that will threaten our free institutions." These words, spoken more than 40 years ago, are even truer today.

As shown in the graph below, oil prices continued to rise further as a result of the Iran–Iraq war in 1979 and 1980, and demand for oil decreased. Unfortunately, as the demand for oil eased, the country did not heed the president's warning. For us at the fledgling Solar Energy Research Institute, however, the President's words were encouragement to strive to fulfill his recommendations, which included reducing oil consumption, developing liquid fuels from biomass, insulating American homes, and developing all forms of solar energy as a major resource.

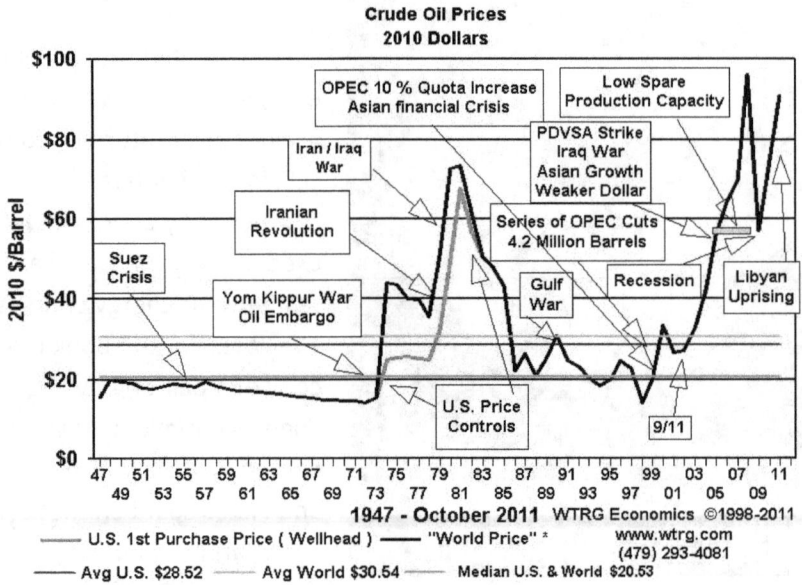

Following the award of the contract to MRI, I reported to work on July 1, 1977, the day SERI officially opened. I deliberately say "reported to work," because I entered an office which had no furniture and the telephone still needed to be hooked up. My badge number was 22, and I surmised that there must be at least 21 other people working at SERI. After having advocated the need for developing solar renewable energy for over 20 years, it was an exciting moment for me to enter a building dedicated solely to advancing solar energy. But the question of how to start when there is nothing to start with loomed large. It did not take very long, however, for me to realize that the first job I needed to tackle was to find the right people to work for the Institute. A task force consisting of Gordon Gross, Charlie Grosskreutz, and myself was formed to screen the many hundreds of applications which had come in from people all over the country anxious to assist in developing solar energy. The choice was not easy because so many superbly qualified people applied.

A week after I moved into my new office, I received a call from a newspaper reporter at the *Denver Post*. After some cursory introduction and inquiries about the general nature of SERI, he asked me, "And what would you, Dr. Kreith, consider to be the most significant advance made in your new job?" I was taken aback by this unexpected interview so early in the life of the institute, and I can still recall my answer, which was "The telephone in my office has been connected and a ream of stationery with SERI letterhead was delivered this morning to my office. In addition we are just beginning to look through a pile of 100 or more applications for employment." Needless to say, the reporter realized that his effort to achieve a coup by being the first to report progress on solar energy at SERI had been somewhat premature.

The first director of SERI was Dr. Paul Rappaport, a scientist who had made a name for himself as director of the Process and Applied Materials Laboratory at RCA's David Sernoff Research Laboratory with the early development of photovoltaic cells. He had been the director of advanced research at RCA for many years, and was an idealistic innovator who did not like bureaucratic restrictions imposed by the U.S. Department of Energy (DOE) in Washington. We shared a similar approach to research and development, and Paul showed his confidence in me by supporting all my plans for building the world's foremost solar thermal program at SERI.

The Thermal Conversion Branch at the National Solar Energy Research Institute was officially inaugurated on July 7, 1978. I was thrilled to be "on board" as the Chief of the first branch to start operation at the Institute. When the Institute began, its charter was that it would be the "Brain Center" for solar activities in Washington. The instructions from the directors of the Institute to the "Brain's" chiefs were to critically analyze DOE operations and to identify those areas in solar energy conversion and utilization which were in need of a fresh look, additional research, or suitable for major thrusts in the near future. The initial concept of operation of the new Solar Energy Research Institute was to be primarily engaged in in-house research for innovative ideas and be available to give policy guidance to decision makers in Washington on the Contract Research and Development which was to introduce solar energy in the marketplace. In accordance with these instructions, the key members of my branch began to examine the research thrusts, the funding, and the technical opportunities and problems of various solar-thermal options. An important factor in the operation of SERI that I was unaware of at the time was that Congress had mandated the establishment of the institute without providing dedicated funding. As a result DOE was forced to provide the money necessary to operate SERI by reducing funding for other programs under its jurisdiction.

Based on the initial analyses, we began to lay out a program suitable for the Institute. It was recognized that an Institute of the size envisioned by Congress (about 1,000 people by the year 1982) could not undertake large-scale commercial development and would have to direct its activities to the key problems that offered high payback in technical know-how and potential solar energy utilization. In the course of the second week of operation, I dictated a memo to my staff identifying six main areas of emphasis for the Thermal Conversion Branch:

1. Utilization of solar energy for the generation of liquid fuels.

2. Utilization of solar energy for domestic hot water.

3. Research into inexpensive concentrating collector construction.

4. Development of ocean energy conversion systems, especially open cycle thermal energy conversion.

5. Application of solar energy to industrial process heat requirements.

6. Improvement in high temperature receivers for solar-thermal electric systems.

The rationale behind the identification of these key issues was that, in my opinion, the United States faced not so much an "energy crisis," as a looming shortage of cheap and readily transportable energy, particularly liquid fuels, and that use of fossil fuels had to be curtailed There is plenty of solar energy falling on earth, but its utilization was not economical because it is intermittent and could not compete with cheap fossil fuels. To reduce the cost of solar energy to the user, I realized that it would be necessary to find applications where the conversion systems and/or collectors could be used all year round at high efficiency. It was also clear to me that greatest needs for energy were in places where solar energy was least available and, therefore, storage and transport of solar energy from areas of high winds and insolation to areas with scarce solar resources was of paramount importance for large-scale development These priorities are of course still with us.

In response to a request from Mr. J.J. Yacovella of MRI to provide a description of the goals of the Thermal Conversion Branch for the organizational directory, I sent him the following memorandum on November 22, 1978, which officially outlined the research goals for the branch:

The goal of the Thermal Conversion Branch at SERI is to conduct and direct research, development, and analysis of solar thermal conversion components and systems necessary to achieve technical and economic viability for utilization of solar energy on a large scale. The major thrusts of the branch are:

Industrial Process Heat—Industrial Process Heat is one of the most promising near-term application of solar energy. Research and development in this area is directed towards improving the performance of intermediate temperature thermal conversion systems as well as to reducing the cost of heat delivered from them.

Solar Thermal Research and Development—the major goal of this program is to develop high temperature solar thermal conversion devices and inter-relate individual components into working systems which can provide solar heat in the upper temperature range for electric power generation. Development includes basic research in heat and mass transfer processes, as well as of better materials for receivers and reflectors.

Ocean Energy Systems—Solar energy stored in the temperature gradients of the ocean will be investigated with an effort to develop systems that can utilize the temperature difference to produce electric energy, and the kinetic energy in the ocean waves to provide power for other types of prime movers. Also, basic research in direct contact heat exchange processes which are integral to the success of Ocean Thermal Energy Conversion (OTEC) systems will be studied.

Thermal Storage—Research and development in this area will be directed toward development of storage devices and mechanisms which can provide solar energy to places with low availability of solar resources. Emphasis will be placed on chemical-reversible reactions which can be driven by solar energy to generate fuels that can be stored and utilized to generate heat at another time and place.

Desiccant Cooling—Basic research and systems study in the area of desiccant cooling are to be conducted in an effort to ascertain whether or not this means of utilizing solar energy can become economically and technically viable.

Liquid Fuels—Research and development to produce liquid fuels from cellulosic biomass, to replace gasoline and diesel fuel without competing with food production.

The need to produce liquid fuels from cellulosic biomass rather than from the kernels of corn, which was then (and still is) the current technology, was one of the key recommendation of Dr. Charlie Wyman, the Chief of the Chemical Engineering Research section in the branch. This was not an area of my own expertise, but I quickly realized that the ideas of Dr. Wyman were of paramount importance for solving the looming energy crisis described by President Carter. Dr. Wyman had previously demonstrated that the amount of cellulosic biomass available for liquid fuels was sufficiently large to make a significant contribution to future liquid fuel needs without interfering with the utilization of agricultural land for food production. Unfortunately, as was so often the case, these arguments fell on deaf ears in Washington and the budget for cellulosic production of liquid fuels was miniscule in comparison with other areas of research.

About ten months after SERI formally initiated its operation, President Jimmy Carter visited the institute in May of 1978 and dedicated a permanent site for SERI on South Table Mountain in Golden, Colorado. This site had been donated as the permanent site for the Institute in the original proposal. For President Carter's visit, we moved some of the equipment up to the mountain, although we realized already at that time that the top of the mountain, although receiving a lot of sunshine, was not a suitable permanent site because of its inaccessibility. The top of the mountain had a rough road which had been used by police in practicing high speed chases, but had no natural gas access or electric transmission lines. I had been asked to get some of the solar thermal technologies ready for the visit and was able to put one of the Omnium-G dual access tracking dishes into operation. In addition, we had a solar single access tracking trough and several flat plate solar collectors to show. The day had been announced as "Sun Day," but unfortunately the sun was uncooperative and the day was overcast with a little drizzle. All the same, we were all enthused by having the President of the United States come to visit the fledgling laboratory and I had the great pleasure of escorting the President through the solar thermal system on exhibit.

I still remember President Carter asking, "How many acres would a solar thermal plant equivalent to a 1 megawatt nuclear power plant occupy?" This was a very reasonable question, but nobody in the president's entourage was able to answer it on the spot. I was quite embarrassed and recall Mike Noland, the Deputy Executive Director for Operations at the time, telling the president that SERI would send him a response within 24 hours.

On May 15th, Paul Rappaport issued a memo informing the staff of reorganization. SERI was to be divided into three groups with technology development under Dr. Charlie Grosskreutz, technology dissemination under George Warfield, and administration under Don Burke. At the time, I did not understand why there should be a reorganization, but in retrospect I believe that the main purpose was to coordinate all of MRI's solar responsibility into every element of SERI, and thereby essentially control the activities at the Institute. This structure for the MRI/SERI coordinating board included, in addition to some of the SERI personnel, Dr. Harold Hubbard a longtime employee at MRI, who subsequently became the director of SERI. The reorganization did not affect me personally. The Solar Thermal Conversion Branch remained essentially intact, but was later renamed the Solar Thermal Research Branch. Also, it was now required to report through an Assistant Director of Research to the Director for Technology Development, Dr. Charlie Grosskreutz.

One of the first tasks I assigned to the team assembled for my branch was to identify the most promising near- and medium-term technologies for utilizing solar energy. The result of this analysis identified applications dealing with solar industrial process heat as more promising than heating of buildings, a conclusion supporting our previous assessment favoring domestic hot water heating. In contrast to heating of a building, where the solar installation could only be utilized during the winter months, industrial and domestic hot water applications could use the solar conversion hardware all year round, and therefore represented a more energy efficient and more economical approach for using solar energy. After having identified this technology, we set to work in designing and building test facilities that could assist in the R&D of solar industrial process heat. An offshoot of this particular technology was of course also the generation of electricity, which can be produced by various engines once a working fluid has been elevated to a temperature where the temperature difference between the solar-heated fluid and the environment can yield a sufficiently high thermodynamic efficiency.

The first two-and-a-half years at SERI were filled with some of the most exciting and rewarding professional and personal experiences of my life. In 1977, a revised edition of the first solar book that I had written the previous year with Jan Kreider, entitled *Solar Heating and Cooling*, was published by the Hemisphere Publishing Corporation. In 1978, the first edition of *Principles of Solar Engineering*, also co-authored with Jan Kreider, was published by McGraw-Hill and adopted as a text by many universities teaching courses in solar energy.

In 1978, I also began preparation for a new peer-reviewed journal for the solar energy community under the auspices of ASME. With the help of my friend Dr. Steve Sargent, we were able to persuade the editorial department of ASME to publish this new journal, called the *Journal of Solar Energy Engineering*. I was appointed its Founding Editor and the first issue came out in February of 1980. The cover, including the list of associate editors is shown below.

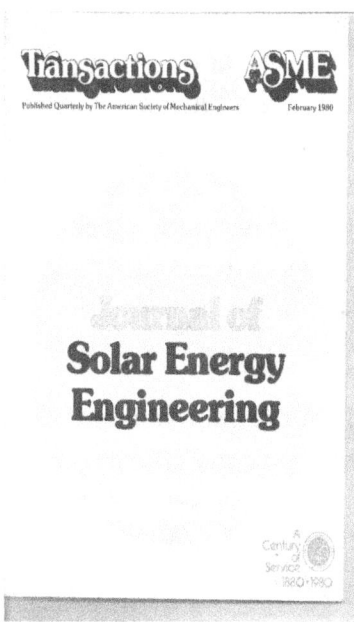

Transactions ASME

Published Quarterly by The American Society of Mechanical Engineers   February 1980

Journal of

Solar Energy Engineering

A
Century
of
Service
1880·1980

ASME Readies Solar Engineering Journal

Meeting the need for better communication within the solar energy community, ASME's Policy Board, Communications, recently voted to establish a *Transactions Journal of Solar Energy Engineering*. Frank Kreith of the Solar Energy Research Institute and the University of Colorado will be the technical editor.

The topics which the journal will cover, and the associate editor for each topic are as follows:

• Testing and simulation — Peter Jenkins, Texas A&M University
• Fundamental processes and theory — K. G. T. Hollands, University of Waterloo
• Heating and cooling — H. M. Curran, Hittman Associates
• Optimization and control — C. Byron Winn, Solar Environmental Engineering Co.
• Industrial process heat — Steve Sargent, DOE/SERI
• Solar electric power — John Gintz, Boeing Engineering and Construction
• Energy storage — C. J. Swet, consultant
• Ocean energy systems — G. L. Dugger, Johns Hopkins University
• Solar collector technology — Vern Rees, Suntec Systems
• Wind energy conversion — R. Oman, Grumman Aerospace Corp.
• Codes and standards — A. J. Baldwin, Honeywell, Inc.
• Materials and structures — to be appointed

The objective of the journal is to present research papers and to translate the current state of the art of solar engineering R&D into a form suitable for practical engineering design. Papers that report on the materials, structures, and the development of standards of solar conversion systems will be included. Emphasis will be placed on design and on the economic aspects of developing solar energy.

In his proposal for the new journal, Kreith pointed out that the federal R&D budget for solar energy is expected to rise from $560 million in fiscal 1979 to $646 million in fiscal 1980; and the number of people working in solar engineering is expected to increase accordingly. By launching this journal, ASME hopes to establish a leading role in this resurgent field.

The review procedure for the journal will be similar to the current one for Division papers. The sources for obtaining papers are: ASME meetings, unsolicited papers, papers presented at DOE-sponsored workshops and conferences, and invited reviews and state-of-the-art summaries.

For the fiscal year 1979–80, ASME members can order the first three issues of the *Journal of Solar Energy Engineering* at the special price of $18 (U.S. and Canada) and $19 (all others). The regular annual subscription price is $50; $25 for members. The price for all 13 *Transactions* will be $560; $280 for members.

The euphoria of starting the first U.S. peer-reviewed journal in the field of solar engineering waned pretty soon after I began to understand some of the details involved in editing a technical journal. First of all, there was no budget for an editorial assistant. It was very fortunate that Marion had the skills to help and was also willing to offer her assistance without remuneration. Working with her as a team certainly helped to cement our marriage and there were compensations other than financial in publishing a journal in a popular field such as solar energy. However, the popularity of solar energy was most intense among our friends and it turned out that the lack of funding for research to academic institutions in the solar field severely limited the amount of publishable research available for the journal. We were able to recruit an outstanding list of associate editors who also began searching for articles that could pass a rigorous peer review necessary for the journal to achieve international recognition. Marion enjoyed the contact with real and potential authors and her ability to keep track of all the correspondence necessary to recruit potential authors, having them prepare the articles in a style acceptable to that set by the ASME, finding qualified engineers willing to review the articles, passing on the review comments to the authors and then making sure that the authors would meet the requirements for publication in the journal, was amazing. This turned out to be quite a lot of work and Marion handled it with cheerful attention to our goal. Mind you, this was before the advent of the Internet and all the material had to be transmitted by snail mail! Marion became so well known at the local post office that the clerk told her," Ordinarily we ask for two IDs if one pays by check, but you only need to show me one."

In addition to the indirect interchange with solar engineers, we also had the opportunity of meeting many of them in person at professional meetings and congresses where we could also meet and talk to the associate editors and the authors of the articles for the journal. Many of the authors and associate editors became our personal friends and the solar community duly appreciated the effort we put into providing a sounding board for the emerging topic of sustainability. There were several times when we barely had enough articles to justify the quarterly publication of the journal necessary for its survival, but it did survive for the first ten years under our guidance. Finally, the number of articles in the solar field began to increase during the period that I was the journal editor, so that when my term ended we had a considerable backlog to turn over to the next editor.

On September 8, 1978, I presented an invited lecture at the International Solar Conference in Haifa, Israel and was able to inspect the domestic solar hot water heaters which were beginning to penetrate the market there. In November of 1978, I was appointed one of the technical experts for the NATO program of Science for Stability and represented NATO in that capacity in Turkey. In the summer of 1979, I had many exciting opportunities to speak about solar energy all over the world. In June, I presented lectures in Stuttgart and Freiburg, Germany, on the potential of solar energy for the International Communication Agency. In July, I was appointed as the U.S. Technical Representative to the United Nations Economic Commission for Europe on solar energy utilization held in the Canary Islands of Spain. This was an exciting and unforgettable experience.

Marion and I traveled to Spain expecting to take a boat to the Canary Islands, but were very surprised to find that these islands were many hundreds of miles from the mainland. Visiting the Canary Islands was a wonderful experience, even though we were shocked to learn that none of the natives that were living on the islands when the Europeans arrived were still alive. One of the last of the natives found living on these islands was taken in chains some 300 years earlier as a slave to the Pope in Rome and died there.

One of the original advances in concentrating solar energy was a furnace built by

French Engineers at Odeillo in the beautiful Pyrenees mountains in the early 1960s. On our journey from France to Spain, Marion and I decided to take a side trip to look at the Odeillo furnace. This furnace uses hundreds of tracking mirrors to concentrate solar radiation from the sun, which is at about 6,000 Kelvin, onto a spherical mirror and from there onto a small spot at the focal point where a temperature of approximately 4,000 Kelvin could be achieved. The furnace was not intended for power production, but rather as a research tool to melt various materials for industrial purposes. Irrespective of its original goal, the furnace demonstrated the ability of high concentration–high temperature solar concentrator technology.

A most remarkable event of this meeting in the Canary Islands was a reunion between engineers who had fought on opposite sides of the Spanish Civil War in the late 1930s. Some of the engineers were, at that time, young people fighting for the fascist army of General Franco, while some others fought for the democratically elected Republican government. The armies of Franco were supported by the fascist governments of Germany and Italy. Support for the armies fighting for the socialist Republican government came only from the Soviet Union, while the democratic governments of Europe and the United States declared neutrality and refused to sell weapons to the socialists. The government forces were, however, supported by what was known as the International Brigade, a collection of volunteers from all over the world that fought against the fascists in Spain out of personal conviction. The father of my solar colleague, Steve Sargent, was one of them and brought us a tape with the songs of that brave army. When the armies of General Franco, with the help of the German air force, crushed the government forces, many of its soldiers fled to France, and some of them went on to the Soviet Union. Several of these men subsequently became engineers in the Russian solar energy program, which built a central receiving power tower in the Crimea. The engineers remaining in Spain that were interested in solar energy had done some R&D in solar thermal energy conversion. Some members from each side recognized each other from their school days, fell into each other's arms, and there was general rejoicing. In the aftermath, they asked each other, "How could we have been so stupid to fight one another in a civil war?" I thought at that time that solar energy could one day become a vehicle for peace, a hope that I still carry with me.

Later in the summer, Paul Rappaport asked me to take his place to deliver the plenary lecture at a forthcoming inauguration of the first interdisciplinary research institute in Japan. This institute was located in Tsukuba City, about an hour from Tokyo by one of the country's high speed trains. It was an ultramodern establishment at which experts from various scientific disciplines were given an opportunity to help the country to become a world power in the electronic age. I was glad to take Paul's place, but was surprised by his request since the inauguration was an event of international importance. I was unaware at the time that Paul had been diagnosed with terminal stomach cancer. He died April 21, 1980, soon after I took his place in Japan. Paul had not only been a valuable administrative support for me, but we also became good friends who shared a similar perspective of a solar future.

PAUL RAPPAPORT
1922 - 1980

SERI'S FIRST DIRECTOR

DR. PAUL RAPPAPORT

PHOTOVOLTAIC LABORATORIES

SOLAR ENERGY RESEARCH INSTITUTE

DEDICATED NOVEMBER 20, 1979

IN HONOR OF THE FOUNDING DIRECTOR

CHAPTER 29—SERI UNDER DENNIS HAYES, 1979–1981

It soon became apparent that SERI was not the only organization interested in pursuing research for solar industrial heat. In a memo dated March 14, 1979 from John V. Otts of Sandia National Laboratory to Mr. M. Gutstein of DOE entitled "Review of SERI's Test Facility Plans and Objectives," Mr. Otts recommended, among other things, that "SERI put more emphasis on utilizing Sandia's expertise related to facility planning and operations," and that "Sandia should be responsible for testing, as opposed to SERI's responsibility for research." He also noted that "SERI's temporary facilities and future relocation seems economically problematic." The essence of the three-page memo directed to DOE with copies to some of the SERI administrators was that SERI should keep its nose out of the solar heat conversion activities in existing national laboratories, and should take advantage of the testing facilities in other laboratories and their associated technical expertise. It further suggested that SERI should concentrate on long-term research. Interestingly enough, I was not copied on this memo, although the recommendations of course impacted most heavily on the Thermal Conversion Branch, which was in the process of establishing large-scale testing and research facilities that were in direct conflict with the interests of the existing nuclear and weapons laboratories. Dr. K. Touryan, one of the SERI administrators who were copied on this important memo, passed a blind copy on to me on March 19th without comment. In retrospect, I realize of course that my ignoring these efforts to undercut SERI was a grave mistake. But there was probably little I could have done to change the change the outcome of this power struggle. The deck was stacked in favor of DOE taking control of the solar program and SERI had to put up or shut up.

In looking over my notes from that period, I found a warning I had written to myself that the tasks selected for SERI could also be of interest to other laboratories. The laboratories that were established to develop weapons during the war and nuclear power afterwards were facing budget cuts and were looking for tasks that could receive future support from Washington. I wrote that "SERI being the new kid on the block is likely to face competition." That observation turned out to be true, and sometime early in 1980, I received a visit from an engineer from the Sandia National Laboratory who proposed that the solar thermal activities should be divided up among four laboratories. Flat plate solar collectors, which were close to being commercial, should be the responsibility of Los Alamos; solar trough and power tower technologies should belong to Sandia National Laboratory; solar buildings technology and conservation should be assigned to Lawrence Berkeley; and SERI should take over responsibility for the dual-axis tracking dish technology and high-risk long-term research. I voiced my objection to that division because I had been told that SERI would be the prime solar laboratory and would be able to assign responsibilities rather than be dictated to. This had been the position of Paul Rappaport, and I thought I was stating official policy. However, as often has been true in my life, I was politically naïve. Before taking charge of the solar energy program, the DOE program managers had previously been in the weapons and nuclear area and had long-term relations with the engineers at the other laboratories. They were, therefore, sympathetic to the dilemma of potential layoffs facing these laboratories. Moreover, Paul Rappaport had been diagnosed with stomach cancer in 1979 and had to relinquish his responsibilities as director. When Paul resigned, I lost my most valuable ally and, to make matters worse, President Jimmy Carter twisted the arm of MRI to select as Paul's successor Dennis Hayes, a man who had no credentials as a scientist and whose only claim to fame had been the organization of Sun Day, an event that promulgated solar enthusiasm among young people but without commensurate business and government support. It was rumored that Carter thought Dennis could help his campaign for re-election. In a memo dated July 26, 1979, John McKelvey of MRI announced the appointment of Dennis Hayes to be the executive director of SERI. He further stated that, as an outgrowth of extensive reassessments of current and future SERI needs, different leadership was needed at SERI. This was an indirect slap at Paul Rappaport's leadership, prompted by DOE criticisms. He further stated that "Dennis, with the continuing and able assistance and support of Mike Noland, can provide such leadership." He promised that Dennis and Mike would be working closely with DOE and Congress to secure widespread use of solar technology as a major source for "satisfying our nation's energy needs in the years to come."

This promise of leadership under the joint direction of a politician and a scientist unfortunately proved to be false. Three days after McKelvey's memo, Michael C. Noland, SERI's deputy executive director, a solar engineering expert with a doctorate in mechanical engineering from Georgia Tech, who had been a director of engineering services for MRI since 1977, resigned. He said he felt obliged to leave the organization after Paul Rappaport was fired by DOE. Mike Noland sent a memo to the staff on July 30th saying that, "When the director of a national laboratory is replaced by a political appointee, the laboratory is changed into an organization in which the next tier of management would be expected to resign." This development left SERI at the mercy of a young and inexperienced director. Dennis Hayes, was an amiable young man, but his personality and political orientation immediately incurred the animosity of the DOE establishment in Washington. Moreover, he did not have the respect of the technical experts at the institute.

Realizing that budget cuts and staff reduction were unavoidable due to the Reagan Administration's directive, a group consisting of Dana Moran, Assistant to the Director; Gordon Gross, Chief of Material Research; Ken Tourien, Chief Scientist; Tom Milne, Chief of Biomass; Ben Shelpuck, Manager of the Ocean System Program; and myself wrote a memo to the acting director of SERI, recommending that Dr. Charles Grosskreutz, SERI's first Director of Research be appointed the permanent SERI director. Dr. Grosskreutz had long experience not only in solar thermal conversion, but also in all other aspects of solar energy. We emphasized that Dr. Grosskreutz was able to maintain a healthy attitude and that he had gained the confidence and esteem of the research staff. Unfortunately, our recommendations to have a competent solar scientist become the director were ignored and Dennis Hayes was selected by MRI to head SERI in the new direction. In retrospect, it is of course apparent that MRI had little interest in solar energy per se, but had its eye mainly on continuing to receive the percentage as fee for its management of SERI.

There was one topic in the renewable energy field in which the Solar Energy Research Institute did not face competition from another national laboratory. This topic was ocean thermal energy conversion (OTEC). The temperature difference between the surface and deep water in the ocean represents a vast potential resource, but one of the basic tenets of thermodynamics, enunciated by Carnot, a 22-year-old French artillery officer more than 200 years ago, is that the highest efficiency any thermal power plant can achieve is equal to the temperature difference between the high temperature source and the low temperature sink, Δt, divided by the absolute temperature source, t, or:

$$Efficiency = \frac{\Delta t}{t_{source, absolute}}$$

The problem with OTEC is that the available temperature difference is relatively small and, therefore, the efficiency of the system relatively low. The concept of utilizing ocean solar thermal energy was first described the French engineer d'Arsoval in 1891, and a prototype power plant was built and operated in 1930 by G. Claude off Cuba. OTEC is attractive because the temperature difference is available 24-hours per day throughout the year, and the potential energy is enormous. However, because of the low efficiency, the heat exchanger to capture the energy is very large and dirt build-up on the surface from the ocean can diminish the efficiency with time. As a result, this large resource has not attracted as much attention as using terrestrial solar energy directly. However, at the time SERI became interested in ocean energy, a new concept had been developed that gave renewed hope to OTEC. The idea was to eliminate the fouling potential of the heat exchanger surface by using direct contact heat exchange between the surface layer heated by the sun and the cooler ocean water used as the working fluid for the turbine. The concept was attractive because one of the members of the branch, the Italian physicist, Dr. Federica Zangrando, had experience in the motion of fluid under a temperature gradient, and Dr. Desikan Barathan, a senior SERI engineer, as well as I, had considerable experience in heat transfer between fluids. We applied for support to DOE and, after receiving a modest grant, we preceded to build a small experimental facility at the Institute in Golden. DOE named one of the most experienced SERI engineers in my branch, Mr. Ben Shelpuck, as the OTEC project manager. When the initial experimental results at SERI were encouraging, Dr. Zangrando was asked to build a larger facility in the coastal waters of Hawaii, where a relatively large temperature difference between top and bottom ocean layers exists. The idea continued to interest me and when I received the Max Jakob Heat Transfer Award in 1986 I used the experimental and analytical work on direct contact heat transfer conducted at SERI as the basis for my lecture which was subsequently published in the ASME Transactions.

SERI continued to do some of the more fundamental experiments to evaluate various physical configurations for exchanging heat between fluids by direct contact. One of these configurations was a structured packing. For this invention Dr. Barathan and I received the Energy Efficiency Technology Award from a consortium of engineering societies in 1989 at the Energy-sources Technology Conference. The citation read: For Advancing Mechanical Engineering in Energy Resource Related Technology with "Direct Contact Heat Transfer with Structured Packing."

Before deciding to proceed with a full scale system, DOE appointed a panel of experts to evaluate the economic feasibility of OTEC power. The panel concluded that, at the present state of the art, even with direct contact heat transfer that eliminated much of the cost of a heat exchanger, OTEC would not be a competitive technology. After the negative economic evaluation, the experimental facilities at SERI and in Hawaii received no more funding and the work had to be discontinued.

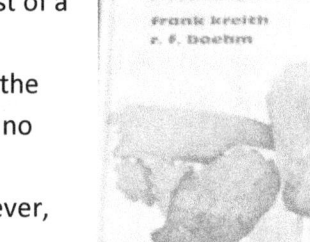

Key Advantages of the NREL-Designed Advanced Direct-Contact Condensers (relative to conventional direct-contact condensers):

- Achieves a lower condenser pressure to boost power plant efficiency

- Operates at a higher thermal efficiency

- Experiences minimal performance degradation by noncondensible gases

- Less susceptible to fouling and corrosion

- Requires less cooling water

- Results in lower hydrogen sulfide abatement costs

- Uses less energy for gas ejectors and vacuum pumps

- Requires less volume, with a reduced need for auxiliary support structures and piping.

To create a better packing for geothermal condensers, NREL researchers modeled and tested a variety of structures. Credit: NREL

However, other experts recognized the potential importance of using the structured packing arrangement that we developed at SERI for industrial applications and the concept received the Energy Resource Technology Industry award for advancing mechanical engineering in energy-related technology by the Consortium of ASME Advanced Energy Systems, the Ocean Energy, and the Petroleum divisions in 1989. A year later, it was awarded one of the IR 100 awards, and Professor R.F. Boehm edited with me a book entitled *Direct Contact Heat Transfer* in 1988 to summarize the state of knowledge for future applications.

Many years later, my friend Desikan Barathan sent me a copy of the NREL magazine, *Innovation*. The main feature of this issue was an article entitled, "Advanced condenser boosts geothermal power plant output." It reported that in the late 1990s, the pressure of the geothermal steam at the geysers was falling and reduced the output of the geothermal power plant because it decreased the temperature difference available for the system. When power productions at the Geyser Geothermal Complex began to falter, NREL engineers designed an advanced condenser based upon the results obtained from the OTEC experiments conducted in the 1980s that dramatically boosted production. After the installation of the packed direct contact condenser mechanism, the efficiency of the plant returned to its previous value.

This instance of the application of a concept developed at NREL and applied in the real world was cited as a case study illustrating NREL's contribution in market-relevant research through its development work. It was certainly gratifying to learn that the R&D Desikan and I conducted in the late 1980s has eventually found such an important application 20 years later in the renewable energy field.

The negative DOE evaluation of ocean thermal energy technology in the 1980s was reversed by the large American defense contractor Lockheed, who signed a contract with the Chinese Beijing-based Reignwood Group in 2013 to design and build a 10 Megawatt OTEC plant. The initial stage of this 3.5 year effort was announced by Reuters on October 30, 2013. According to Dan Heller, Vice President of New Ventura at Lockheed, this plant was designed to prove the concept before building a 100 Megawatt ocean thermal power plant for the world market.

Lockheed signs deal to design largest ocean thermal electric plant

BY DAVID ALEXANDER
WASHINGTON   Wed Oct 30, 2013 2:46pm EDT

0 COMMENTS   Tweet 3   Share 1   Share this 8+1 0   Email   Print

Oct 30 (Reuters) - Leading U.S. defense contractor Lockheed Martin signed a contract on Wednesday to design the biggest power station fueled by differences in ocean temperatures, a 10-megawatt plant that would provide electricity for a new Asian resort.

The contract between Lockheed and Beijing-based Reignwood Group, a Chinese consumer products and lifestyle firm, is the initial 10-month stage in a 3-1/2-year effort to build the green energy electric plant, which would generate power using a process known as ocean thermal energy conversion (OTEC).

"This is just more or less the tip of the iceberg and what both parties are most interested in is ultimately getting the plant built so we can offer it to other customers. And that's where the business is for Lockheed," said Dan Heller, vice president of new ventures for Lockheed's Mission Systems and Training unit.

A major blow struck SERI on August 1, 1979. On that day, Charlie Grosskreutz submitted his resignation to Dennis Hayes, citing frustrations and political aspects of his job which made it impossible to carry out his management responsibilities. He said that SERI management "has been unable to establish a working relationship with the DOE Assistant Secretary for Energy Technology to whom SERI reports. During the past year alone, four men had occupied this position: Eric Willis, Bob Thorn, John Deutsch, and Don Kerr, three in an acting capacity. None remained in office long enough to formulate, approve, or sustain decisions on SERI's role and mission." With Grosskreutz's departure, the staff at SERI became even more dependent on its new director, Dennis Hayes, whose lack of technical knowledge and personal arrogance made it difficult to establish confidence in his administration. I am not sure what all the reasons for Charlie Grosskreutz's departure were, but according to an article in *Science* dated March 23, 1979, a policy researcher said, "The place is pandemonium. Nobody on the left hand knows what the right hand is doing. If it were a private firm, it would have been broke ages ago." A second person said, "At SERI, the Peter Principle has become an epidemic. The great majority of people in management have no qualifications to hold the positions they hold." The article further stated that "projects have started at SERI and are stopped and revised repeatedly by headquarters. This has produced terrible animosity at times with DOE and SERI officials swearing at one another in conferences. Even Rappaport at some point had conceded that some of the brilliant people on his staff occasionally must kowtow to less knowledgeable counterparts at DOE. The DOE manager, Bennett Miller, on the other hand complained that, 'If only SERI would learn to play the game, at least half of each assignment from DOE could be used to pursue ideas that SERI likes.'" Obviously, this continuing disagreement between SERI and DOE on how to run the solar program made life difficult for both organizations and delayed commercial applications of many promising ideas.

Soon after Reagan was elected president, Hayes blasted the DOE and the Reagan Administration, accusing them of "open war on solar energy." When John McKelvey, the Director of MRI, learned of this comment he promptly ordered Hayes to leave his post immediately. It was announced that Donald Feucht would be the acting director until Harold Hubbard, a vice-president of MRI, could take over on July 1st. Hubbard was scheduled to serve until a permanent director could be found. On June 30th, the acting director sent a memo to SERI's staff regarding the SERI transition status. He said that the plan for realigning SERI to a long-range laboratory-oriented research and development mission consistent with the revised DOE solar energy goals and budget would start immediately. He stated that, consistent with the new direction, the following programs were to be terminated: public and consumer education programs, socioeconomic studies, international activities, information dissemination to the general public and subcontracting activities not in direct support of research. Furthermore, he stated that, "…… consistent with these program changes, MRI will make the necessary reductions in SERI staff."

Hayes' firing came only a few days after MRI revealed that in response to the Reagan Administration's request, a cut in SERI's budget from $120 million to $50 million per year was necessary, and this cut would eliminate programs that did not involve long range technical goals. This included the program to disseminate information about the potential of renewable energy. The new budget proposal would eliminate some 350 of SERI's 950 positions. At the last moment, Colorado senator, Gary Hart, tried unsuccessfully to add $450 million to the 1982 budget. He said that SERI's demonstration and commercialization programs, which were to be eliminated, had proven successes. He said, "This country cannot afford policy reversals that will further delay the date on which we can declare our independence on foreign sources of oil."The cutback in funding forced me to make some painful decisions. I had recruited some of the most talented engineers and scientists who were committed to developing solar energy as a means of solving the energy crisis and promoting world peace. Now, 2 years later, I had to tell them that there was no funding for their position in next year's budget. As a result most of the best researchers began to look for other positions.

In the summer of 1979, I also had to make a crucial personal decision. I had started my work at SERI on a two-year leave of absence from the University of Colorado, which was to expire in August of 1979. My chairman, Dr. Fred Ramirez, told me that it was not permitted under university regulations to give another leave and I would have to resign my tenured professorship in the department of Chemical Engineering unless I returned to teaching at CU in September. It was a very difficult decision, because I knew that a position at SERI was not as secure as a tenured professorship. Moreover, Paul Rappaport had resigned a few months before (in July of 1979) and I knew very little about his replacement, Dennis Hayes at the time. My decision was greatly influenced by reading an autobiography by Golda Meir, the Prime Minister of Israel who had emigrated from Colorado to Israel in order to help the new state grow. In this book she said something like "Few people have the opportunity to decide on a course of action towards a goal in which they truly believe; if you ever have such a chance grab it, even if it means taking a risk into an unknown future." I took those words to heart and wrote a letter resigning from the university. Thus I became a Professor Emeritus in 1980, twenty years after coming to the University of Colorado. Fortunately, this was not the complete termination of my association with the university.

CHAPTER 30—SERI'S FATE UNDER RONALD REAGAN, 1980–1983

Despite the ominous clouds on the horizon, the Solar Thermal Research Branch, which was renamed the Thermal Conversion Branch in March of 1980, continued to prosper for a while longer. Since the budget for the year 1980 had been approved there were still ample funds for continuing the R&D of the branch and achieve many of the goals which were initially set. The zenith of my career at SERI occurred in the fall of 1980, when I was Chief of the Solar Thermal Research Branch and Acting Group Leader of the Innovative Concept and Analysis Group. The branch had grown to about 40 people, some of whom are shown in the picture.

The annual payroll was about $1 million and the total operating budget about $5.7 million. I chaired the Industrial Process Heat Committee of the American section of the International Solar Energy Society (ISES), was elected to the executive division of the American section of ISES, served as a member of the executive committee of the Solar

Energy Division of ASME, and was elected to the executive committee of the Solar Radiation section of ISES. I also served on Colorado Governor Richard Lamm's solar energy advisory committee and was asked by NATO to assist in allocating funds for the Science of Stability program in the field of solar energy. The director of the Archives at the University of Wyoming invited me to send all my technical material for safekeeping.

1920-21

HONE 307-766-4114

THE UNIVERSITY OF WYOMING
ARCHIVE OF CONTEMPORARY HISTORY
BOX 3334
LARAMIE, WYOMING 82071
June 16, 1981

Mr. Frank Kreith, Branch Chief
Solar Thermal Research
Solar Energy Research Institute
1617 Cole Boulevard
Golden, CO 8040;

Dear Mr. Kreith:

Thank you very much for your wonderful letter of June first, and for the material that you sent along for your collection. I can assure you, without hesitation or qualification, all the items you enclosed were of interest to us. We would be especially grateful to you if you would be willing to add to the collection per your convenience. Indeed, as I probably mentioned before, when we ask for a person's papers, in effect, what we are doing is requesting a biography in documents. Like any biography, the more complete it is, the more valuable, historically speaking. So under this rather all-inclusive archival premise, almost anything you are going to have relating to your career in the field of solar energy would be of special interest to us and valued by us.

Thank you so much for writing, and for contributing. I hope that we can keep in touch.

With all good wishes,

Cordially,

Gene M. Gressley
Director

RECEIVED
FRANK KREITH

GMG/jmr

JUN 2 2 1981

Also, as previously mentioned, I was the technical editor of the ASME Transactions *Journal of Solar Energy Engineering*, which had started operation a year earlier. The journal was beginning to prosper with the help of Marion as the secretary and the many associate editors who provided support by reviewing articles submitted for publication.

Among the goals that I set for the branch was to achieve an international reputation of excellence, contribute to the knowledge of solar thermal conversion, communicate that knowledge through professional papers and technical presentations, and finally, maybe most importantly, to remove economic and technical barriers that hindered commercialization of solar thermal conversion. As I later learned to my chagrin, missing in those goals was to obtain financial support from DOE and other government organizations, establish close coordination with the solar establishment in Washington, and seek new sources of funding. These omissions undoubtedly contributed to the dissatisfaction of the upper management at SERI with my performance as the branch manager, and to my eventual replacement by Dr. David Johnson, who was the second-in-command at the time. In retrospect, I realize that my real interest was in conducting R&D in the solar field and then disseminating technical information about solar energy on the assumption that this information would then lead to substantial market penetration of the technology. This was in contrast to the primary objective of MRI and its carefully selected Institute administrators, which was to obtain funding from Washington for MRI to continue managing the Solar Energy Research Institute, and thus to collect its 5% management fee.

I believe that the start-up period for any new organization is so exciting because it is a time of building and full of hope for the future. During that period, personal animosities and personality differences are held in abeyance because the common goal overshadows such potential problems. Unfortunately, this period slowly comes to an end when the organization stabilizes and develops a bureaucracy that emphasizes goals different from those in the idealistic beginning. The people who succeed in a bureaucracy have a more practical and political personality, and their skills are quite different from those of the researchers and innovators who start organizations. This type of transition unfortunately also happened at SERI and was exacerbated by political changes in Washington.

In 1980, I was invited by the Sigma Xi honorary science society to become one of its national lecturers. I selected as the topic of my lecture "Solar Energy: Promise and Reality." With the interest in solar energy revived under President Carter, my lecture topic was of wide interest and I presented Sigma Xi lectures in the following two years at the University of Idaho, the University of Northern Arizona, the California Polytechnic Institute, Minot State College in North Dakota, the University of Montana, the Mayo Clinic in Rochester, Minnesota, the University of Toronto, Canada, and the Marathon Oil Company.

Frank Kreith
Solar Energy Research Institute
1536 Cole Boulevard
Golden, CO 80401

Industrial Application of Solar Energy (S)

Solar Energy: Promise and Reality (G)

Thermal Performance of Solar Systems (N)

Dr. Kreith currently serves as chief of the Thermal Conversion Branch, Solar Energy Research Institute, while on leave from his position as Professor of Engineering at the University of Colorado. He holds degrees in mechanical engineering from the University of California–Berkeley and a doctorate from the University of Paris. He was a research engineer at the Jet Propulsion Laboratory at California Institute of Technology and a Guggenheim Fellow at Princeton; he has served on the faculties of the University of California–Berkeley, and Lehigh University. Dr. Kreith has also held Fulbright and NATO Senior Fellowships.

Sigma Xi National Lecturers 1980–81

For the 43rd successive year, Sigma Xi has assembled its College of National Lecturers, in order to give chapters and clubs the opportunity to hear nationally known scientists discuss particularly lively areas of current research in a manner suitable for interdisciplinary audiences.

The outstanding scientists listed here have agreed to make themselves available to speak to any Sigma Xi group insofar as their commitments permit. They will serve from July 1980 through June 1981 and have consented, in support of the Society's objectives, to limit their honoraria to a minimal $150 per lecture, together with full payment of travel costs and subsistence (provided either through hospitality or reimbursement).

Invitations should be sent directly to the lecturer, and all arrangements about dates, travel, and hospitality should be worked out with the individual lecturer. Two important points: once these arrangements are firm, it is important to inform the chairman of the Committee on Lectureships of them; and following the visit, the lecturer should promptly be given the honorarium and reimbursement for out-of-pocket expenses.

The Society has instituted a modest program of subsidies for chapters and clubs in need of assistance in financing the visit of a National Lecturer. In awarding these funds the Committee on Lectureships gives priority to chapters and clubs that, because of their small size or remote location, might have difficulty in attracting an outstanding scientist to visit them and that have shown initiative in obtaining local resources to partially support the cost of a National Lecturer's visit. Applications to cover visits during the 1980–81 academic year will be due in the Office of the Committee on Lectureships by 1 May 1980. In January an application form was sent to the president of each chapter and club. It must be emphasized that this is an application for *subsidy only;* invitations and all arrangements are to be worked out between the chapter or club and lecturer directly.

To assist chapter and club officers in planning their programs, we have asked this year's lecturers to designate the appropriateness of their lectures for different types of audiences, as follows:

S Sigma Xi audience of scientists from a broad range of disciplines

G General audience including high school science students and/or nonscientist guests

N Specialized audience representing disciplines close to that of lecturer

John W. Prados, Chairman
Committee on Lectureships

I enjoyed these Sigma Xi lectures enormously because they gave me the opportunity to interact with people who were interested in solar energy, but had no previous technical knowledge about the promise and limitations of this unlimited energy source. In retrospect, it seems however as though I was a lone voice howling in the desert of an energy industry interested merely in finding more oil to satisfy the increasing demands of society. And although the lectures may have raised awareness of solar energy and brought prestige to SERI, they brought no outside funding to the institute.

The Iranian hostage crisis created political problems for President Carter in his 1979 bid for re-election, and he subsequently lost the presidential election to Ronald Reagan in 1980. After he was elected, President Reagan claimed that "Alaska has a greater oil reserve than Saudi Arabia." Reagan did not believe in a solar energy future and one of his first steps after taking office was to remove the solar collectors that President Carter had installed on the roof of the White House. This was not merely a change in physical appearance, but also had enormous symbolic significance. I believed at the time that President Reagan had missed the truly great opportunity to provide for a secure energy future for the country and I wrote in my notebook " ... this act will delay our renewable solar energy future, but it cannot avoid the eventual transition from our fossil fuel dependent energy system." According to Leslie Clark of the McClatchy

Tribune's Washington bureau (May 9, 2014), in response to prodding by environmentalists, the then-Energy Secretary Steven Chu pledged that solar panels and a solar water heater would be installed on the historic building roof of the White House. President Obama announced the completion of the installation on May 9, 2014.

There were many other actions taken by President Reagan that held back the momentum for sustainable energy development started under President Carter. I am still amazed why Reagan is held in such high esteem today. But for me the situation offered some new challenges.

In an effort to provide information about economics and technologies of solar energy and energy conservation to a broad audience, my friend Ron West embarked

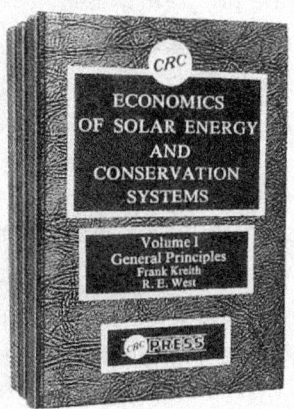

with me on our first major publication venture. It was a three-volume handbook on the technology and economics of renewable energy and conservation, published by CRC in 1980.

This work by Ron and me was also the beginning of a continuing association with the book

Available Now

ECONOMICS of SOLAR ENERGY and CONSERVATION SYSTEMS

By FRANK KREITH, D.Sc., P.E., Chief, Thermal Conversion Branch, Solar Energy Research Institute, Golden, Colorado, and Professor of Chemical Engineering, University of Colorado, and RONALD E. WEST, Ph.D., P.E., Professor of Chemical Engineering, University of Colorado, Boulder.

All of the new energy technologies, including solar energy, are more complex and expensive than the old oil and gas-based technologies. Therefore, energy conservation is increasingly important. Also, a more sophisticated approach to investment decision in energy supply and conservation systems is required. The three volumes of *The Economics of Solar Energy and Conservation Systems* are designed to provide the economic and technical background necessary to make such investment decisions wisely.

publisher and, more importantly, a life-long personal and professional friendship with a wonderful friend. Ron and I had a symbiotic interaction that made it possible to work jointly on many different topics, solar energy, transportation, hydrogen economy, and fracking, among others. I had a tendency of always trying out new and unorthodox ideas, while Ron provided the rational and more cautious perspective that kept me from going off the deep end. The outcome was always a carefully measured article in a professional journal or a publishable book. I am forever grateful for having such a great friend.

The funding for SERI had reached $ 112 million for FY 1980 and the number of employees was over 950 at the beginning of 1981. The budget for the year had also been approved at a reasonable level and operation of R&D continued for a while without major interruptions. But soon thereafter, as previously stated, the budget for SERI was cut drastically by the Reagan Administration and much promising solar R&D had to be terminated.

After the new administration forced the resignation of Dennis Hayes in March of 1981, the very existence of the institute was in jeopardy. MRI reacted to these developments by installing one of its own people, Dr. Hub Hubbard, as the new director of SERI. He had no previous experience in solar energy, but he was an experienced administrator and his primary objective was to mend fences with DOE and save the institute. I gathered that some of the program managers in Washington resented the phenomenal growth of solar thermal activity at SERI, which they considered to come at the expense of their friends at other laboratories. Dr. Hubbard was more successful in placating the program managers in DC than in fighting for the preeminence of SERI.

As part of these ominous developments, the number of positions assigned to SERI in the solar thermal area was cut drastically and a reduction in force (RIF) became necessary. The RIF was to take effect in July of 1981. Once it became known that drastic budget cuts were looming, many of the SERI employees started looking around for other jobs. As would be expected, the most talented and capable of those people were soon able to find employment elsewhere. As a result, not only the number, but also the quality of the staff remaining at SERI suffered from this RIF. I have looked at the subsequent careers of some of the people who were in my branch and I am amazed at the success they achieved after leaving SERI. Following is a sample of these highly qualified people that were lost from SERI :

Dr. Charles Benham and Dr. Mark Bohn jointly founded Rentech, a start-up company which successfully developed technology for producing liquid fuels by the Fischer-Tropsch gas-to-liquid technology and positioned it for commercialization. Dr. Benham was the president and Dr. Bohn the vice-president and chief engineer of the company.

Dr. Charles Wyman devoted most of his efforts after leaving SERI to advancing technology for the biological conversion of cellulosic biomass to ethanol. He held the Ford Motor Company Chair of Environmental Engineering at the University of California at Riverside, where he focused his research on pretreatment and enzymethic hydrolysis. He is also the co-founder and chief development officer for Mascoma

Corporation, a startup company focused on biomass conversion to ethanol.

After Dr. David Kearney left SERI he founded and became President of Kearney and Associates, a company that provides consulting on the commercial developments and project implementation of solar thermal electric systems. He assisted Luz in the development and operation of SEGS, the largest solar electric plant at its time. SEGS has now operated successfully for 35 years in California..

One of the younger most talented engineers in the branch, Dr. Charles F. Kutcher, remained at SERI, but started a PhD program at CU and returned full-time to SERI after successfully completing his doctorate in 2004. He is today a principal engineer and manager of the Thermal Systems Group at the National Renewable Energy Laboratory. However, he has made a name for himself outside his official duties at the laboratory by becoming one of the best-known spokespeople for the need to reduce CO_2 generation and global warming. In 2008, he also taught a course on the topic at CU.

Ms. Frances H. Arnold went on to obtain a PhD at the University of California in 1985 and is today the Dickenson Professor of Chemical Engineering and Biochemistry at the California Institute of Technology. She has become an eminent authority in her field.

Dr. Ari Rabl first returned to Princeton University and later was appointed Professor at the Ecole des Mines in Paris. He published extensively on performance predictions for solar radiation to heat conversion and evaluation of externalities in energy generation. He received the ASME Frank Kreith Energy Award in 2008.

Ken May and Randy Gee left SERI in 1983 to form a company devoted solely to the development of single axis tracking concentrating solar collectors for intermediate and high temperature applications. The company, Industrial Solar Technology, experienced difficult times when solar energy fell into disfavor, but it survived. Their achievements were recognized by the American Solar Energy Society, who awarded them jointly the first Hoyt Clarke Hottel Award for "unique contributions to make solar thermal technology a commercial reality" Ken May is today Chief Engineer for the Spanish energy company Abengoa which has built the largest solar thermal electric power plants in the world. He was named one of the outstanding alumni of the University of Colorado in 2008. Randy Gee retired in 2013 as Chief Engineer of Sky Fuel, a company devoted to commercialization of solar trough technology using a lightweight reflective film instead of glass mirrors to concentrate to sun energy.

There are several other people from SERI who went on to becoming successful in furthering renewable energy commercially after leaving the institute, but I have not kept track of them. However, despite all the ups and downs of the initial group of engineers who started SERI, and the sad experience of being RIF'd by some of them, at the 25th reunion of the SERI all of the former employees were glad to have been a part of this hopeful and exciting period in the birth of solar energy as a major future energy source.

In January of 2005, Dr. Dan Arvizu became the director of the National Renewable Energy Laboratory, a post that he has held ever since. Under his direction, the laboratory aligned its policies more closely with those of any of the administration in power at the time and averted any of the major political discords that had plagued the institute heretofore. The political alignment came at the price of giving up the effort to influence national energy policy. For example, when President George W. Bush proclaimed hydrogen to be the future fuel, NREL redirected some of its efforts to R&D on hydrogen, despite the realization by its senior members that the hydrogen economy was more a political ploy than a realistic assessment of its technical potential. However, the laboratory has experienced substantial growth, both in funding and in new building infrastructure.

CHAPTER 31—SERI SENIOR RESEARCH FELLOW, 1984–1987

In the fall of 1983, the Director of SERI asked me whether I might be interested in giving up my administrative responsibility and accept one of the two newly-created positions of Senior Research Fellow at the Institute. The other Senior Research Fellow, Dr. Arthur Nozik, was an expert in a rather narrow research area dealing with microbiology. He had come from academia, where the National Science Foundation had supported his research for some time before he came to SERI and he was able to transfer his equipment and funding to SERI. In contrast to Dr. Arthur Nozik, who had his own personal funding and laboratory, I had to give up my experimental facilities because they were a part of the branch and were funded on a year-to-year basis from the overall budget of DOE as a line item. This yearly funding had been a continuous battle because I wanted to devote my efforts to some promising long term projects that were not always the priority in DC. To be honest, I do not believe that I was ever cut out to be a bureaucratic administrator, especially with the ever-changing interests in Washington. One incident that was particularly troublesome for me occurred when my branch completed an assignment four months before the end of the contract. When I informed the DC program manager that we had met our goal and suggested to use the remaining funds for a new initiative, the program manager became irate and said "If they find out that you asked for excessive funding last year and I approved the budget, they will automatically reduce your funding to less than you request for next year." I could not accept this reasoning and instructed the branch to initiate work on the new idea, but did not tell the program manager. That is probably what he wanted me to do, but I will never know.

The main change in which business with DOE was conducted after I handed my administrative responsibilities to Dr. Dave Johnson was to align the work done at SERI/NREL with the program organization in Washington. This meant that a specific Program Manager at DOE would decide what research NREL was to conduct the next year and then request a proposal to do the work (with a hint of how much money to ask for), and then provide funds for that year. For the work that he or she wanted done, the DOE manager would then conduct periodic progress reviews of the program. SERI/NREL gave up its original position as the eminent institute that could direct its own research and the responsible people at SERI/NREL had to make a continued effort to keep the DOE managers happy. These efforts included not only meeting technical milestones set in DC, but also finding out what specific interests outside of work attracted the program managers to Colorado. For example, when one of the program managers expressed interest in skiing, the people at SERI would schedule the program review on a Friday morning, and then coincidentally tell the program manager that some of the employees were going to Vail or Aspen on Friday afternoon and could take him or her along. At the ski resort, if someone had a condo, the program manager could stay as a guest. The program manager would also be given a coupon that would allow him or her to buy a lift ticket at a discount. Clearly, all of these amenities were orchestrated in such a way that they were perfectly legal, but also gave the program manager an incentive to continue supporting the SERI efforts, which also provided him or her with a weekend of skiing at minimal cost. Similar efforts were made for those who liked mountain hiking or golf. I do not know how necessary those efforts of personal attention to the program managers were, but was told that this was just part of the culture of "you scratch my back and I'll scratch yours." Although that was not the way I had hoped to see the nation's primary solar research institute operate, since there was nothing illegal about the arrangements, I reluctantly went along with what the managers at NREL considered important for "establishing good relations" with the Washington program managers who controlled the purse strings. As my good friend Dr. Richard Caputo, a former SERI employee, once remarked to me, "Working for DOE teaches one the Golden Rule: GOLD RULES."

I was told that my job as Senior Research Fellow would be that of an ambassador for solar energy, but no official job description came with my appointment. The job description in SERISCOPE of Oct.15, 1983 reads: "The newly created positions of Senior Research Fellow is a recognition of excellence by colleagues at SERI and by peers in the research community; candidates must be nominated by their superiors at SERI and endorsed by their peers ….The Senior Research Fellow appointment is designed to recognize a national and international authority who influences or changes technology in the scientific field. Additionally, a Senior Research Fellow serves as a consultant for the institute and broadly impacts activities in the field of specialty." I was honored by the appointment because I thought that it fitted my professional goals and was given in recognition of my life-long accomplishments in promoting solar energy. A shortcoming of this change in my responsibilities was that no clear source of funding for the Senior Research Fellow was stipulated. But the idea of being a solar ambassador was sufficiently appealing to me not to ask detailed bureaucratic questions. As it turned out, this was a serious mistake in a world where dog eats dog and the bureaucratic details are more important than the technical contributions of an individual and the promises of an administrator who may not stay in the job for long. I remained the Senior Research Fellow for four years, from the fall of 1983 until the end of 1987. In 1983, SERI expressed its appreciation for my contributions to the institute

by awarding me the first General Achievement award. When Dr. Hubbard handed me the award, a beautiful statue of an Indian warrior, he said that the award was richly deserved.

In 1986 I received an additional honor that I had never expected. The American Society of Mechanical Engineers and the American Institute of Chemical Engineers had many years ago established the Max Jakob Award, and some of my mentors, including Ernst Eckert, L.M.K. Boelter, Warren Rohsenow, and Cheng-Lin Tien, had been previous recipients of this award. The award is considered by the thermal research community as the highest recognition in the field of heat transfer. The award was presented to me at the Eighth International Heat Transfer Conference, "for contributions to heat transfer and solar thermal energy conversion during the past four decades." I do not believe that I can give a better description of my elation than the speech I gave upon receiving this prestigious award.

....

MAX JAKOB MEMORIAL AWARD ACCEPTANCE SPEECH

Delivered by Dr. Frank Kreith, Senior Research Fellow of

SERI, at the Eighth International Heat Transfer Conference on

August 18, 1986, in San Francisco, U.S.A.

I want to express my deep appreciation to my peers and colleagues in the thermal sciences for having selected me to receive the 1985 Max Jakob Memorial Award. The professional status of this award is well known and the list of its previous recipients attests to the high recognition it has in the field of heat transfer. For me, however, this award has a special meaning over and above its professional significance because Max Jakob and I have shared similar world views and personal backgrounds in a turbulent period of world history. We were both refugees from Nazi Germany and found in the

United States the fulfillment of our dreams and expectations. Max Jakob and I have shared a high respect for the challenges of research in heat transfer, enjoyed contacts with students from all over the world, and made a major effort to put our ideas in the form of text books to be used by others...

In accepting this award I want to express my belief that it is not given to me alone. It is just 10 years ago that I left my university laboratory in academia to accept an administrative challenge. This challenge was to build the thermal research laboratories and facilities for the Solar Energy Research Institute, an organization that was mandated by the U. S. Congress in 1975 to assure a safe and reliable supply of energy for the future. The change from doing my own research to guiding the research of others has been a very satisfying experience, but my administrative responsibilities also required time and effort that might otherwise have been devoted to my own work. As a result, I do not have a research paper to present at this conference, but I am pleased to note that several of my younger colleagues from SERI do. Over the past ten years, I have had the good fortune of providing technical and administrative guidance to a group of outstanding researchers who have done an excellent job in building up laboratories for research in solar thermal energy conversion, thermal storage, heat transfer in buildings, ocean thermal energy conversion, direct contact heat transfer, desiccant cooling, double diffusive convection, and natural convection. The work of these men and women has brought a reputation for high quality research in the thermal sciences to the Solar Energy Research Institute, and therefore, I believe that they all have a share in this award.

There are, of course, many other people to whom I owe an enormous debt of gratitude for this memorable moment that allows me to stand before you and receive the Max Jakob Memorial Award. I would like to take a moment to mention just a few. First, there is my wife, Marion. She has helped me in many ways, ever since typing the first edition of Principles of Heat Transfer Then there are my students and the co-authors of my recent books and monographs - Jan Kreider and Ron West of the University of Colorado, Bill Black of Georgia Tech, Bob Boehm of the University of Utah, Charles Hoogendoorn of Delft, Ari Rabl of Princeton, and Ren Anderson and Mark Bohn of SERI. Without their help, I could not have completed the works that were written during the last decade as administrator at SERI.

I now want to eulogize Dr. Paul Rappaport, the founding director of SERI, and the person who gave me the opportunity to develop the thermal science laboratories at the institute. Paul came from an industrial background at RCA, but he had a broad vision of research as a means of helping people all over the world achieve a better life and of solar energy research and development, in particular, as a contribution toward world peace. His ideals attracted some of the most promising young researchers to the fledgling institute during its initial years of growth. I remember him telling young researchers that their future will be like climbing a ladder that has more and more rungs as one ascends it. Climbing step by step up this ladder of research is a lifetime job. But since it takes such a long time, there is one step more important than any other: be sure that you lean the ladder against the right wall before you start to climb!

At this moment, I know in my heart that I have leaned my ladder against the right wall and want to take this opportunity to thank all of you who have interacted with me to make this climb an exhilarating experience that has culminated today in this memorable moment of accepting the Max Jakob Award from you.

....

During the first two years as the Senior Research Fellow, I wrote several peer-reviewed technical articles, gave talks about solar energy at various institutions, participated as a reviewer of research grant proposals for DOE and NSF, but did not have any specific assignments or funding under the SERI financial structure. I therefore accepted visiting professorships at the University of California, Santa Barbara, for the spring of 1985, and at the University of California, San Diego, for the spring of 1987. Both of these appointments were approved by SERI and DOE. I obviously thought that they were in keeping with my assignment as solar ambassador, especially since I introduced and presented new courses on solar energy utilization at both of these prestigious institutions. However, unbeknownst to me, there were a number of people in the SERI Administration who considered the position of solar ambassador unnecessary, especially since the position was not a line item in the funding from Washington. That attitude may have been exacerbated by jealousy for the recognition that I received without contributing financially to the Institute by obtaining grants from DOE or other organizations. Naively, I assumed that my professional contributions fitted the job description since they broadly impacted and advanced activities in the field of solar and renewable energy. The first hint that there may have been a problem was an unexpected invitation for lunch by the SERI Assistant Director Don Feucht. During the lunch Don asked about my future plans. I recall him asking" now that you approach your 65th birthday, would you not like to make a trip around the world?" I told him" I want to continue working at SERI because my last year at SERI have been very productive and that, in my opinion, I could continue to make significant contributions to what the institute expected from its Senior Research Fellow."

In the week of July 13, 1987 SERI observed its tenth anniversary. The institute had survived despite the hostility of the Reagan administration towards solar energy, but the anniversary was not an occasion for celebration. The staff level had dropped to less than half of its prime; more than 90% of the staff was still housed in rented space; yearly funding had decreased from its peak of $120 million to less than $60 million and more than half of that sum was spent on procurement functions with industry and academic organizations. Of the three most noteworthy accomplishments mentioned by Dr. Hubbard in his testimony before the House of Representatives Committee on Science, Space, and Technology Subcommittee on July 8, 1987, none came from the thermal conversion field. One of the examples he mentioned as having been translated into a useful technology for industry was the conversion of wood wastes into liquid fuel. This requires an enzyme that can break up cellulosic material in order to liquefy the biomass, a technology that is badly needed, but to this day has not been commercialized. I was one of the 7 staff members recognized by the director for having been at SERI since its inception. But six months later, only six of the original seven little dwarves were left.

On December 8, a week before my 65th birthday, the secretary of the newly appointed Manger of the Thermal Research Division asked me to come to his office saying that the meeting was urgent. The manager was a stranger to me. He was impeccably well dresses and had a face that seemed incapable of smiling. He had been hired from industry a short while before and had no previous solar experience or technical reputation When I entered his office, he first commented " This yellow tie you are wearing is not appropriate for a Senior SERI employee", and then he informed me that the Division had decided to abolish my position. When I asked for a reason, he simply stated that "there is no charge number for you in the financial budget of the Institute beyond the end of the fiscal year 1987". I said to him "I have been at the Institute for more than a decade, have met all of the high expectations for my position and your attitude towards a Senior Research Fellow is nothing short of disrespectful". I then stood up and told him that I would fight this unreasonable decision. He stared at me in disbelief and then asked me to clear out my desk within 24 hours. When I left his office I remembered what one of the employees who lost his job during the first RIF had said to me: "Working at SERI is like the life of a mushroom. One is kept in the dark, fed manure, and finally canned." At that time I did not think that I would one day be one of the mushrooms.

A few days before I learned of my forced retirement from the Solar Energy Research Institute, I had seen a modern sculpture by the young Japanese artist, Yoshiva Yakinawa, and also met the sculptor personally. I liked another of his sculptures that is exhibited below the Boulder Public Library, in the sculpture park next to Boulder Creek and thought it would be very nice to have one of these sculptures in front of the window next to my desk to look at while I was writing. Yakinawa said that he did have two or three sculptures for sale, and I made arrangements to visit him. After my forced retirement from SERI, however, I had second thoughts about buying an expensive piece of art when I no longer had an income as a professional engineer. But then I decided that the action of the bureaucrats would not stop me from having a beautiful sculpture next to my window and I visited the atelier of the artist. There I found two sculptures quite different from the one in the sculpture garden and much more massive than I had anticipated. However, one of them, which was called the guardian, appealed sufficiently to Marion and me to buy it despite the inopportune time for a large financial expenditure. Yakinawa personally brought the sculpture to our patio and we positioned it exactly where I had hoped to have an opportunity to look at it from my desk. The sculpture has been a part of our life ever since and I have never regretted my decision to buy it, despite having been dismissed from my job.

This was not the first time that I had lost a position, but it was an unexpectedly heavy blow. However, I thought that the decision to terminate me must still have to be approved by many of the people that I had brought to the institute and whose careers I had supported while I was branch chief. I therefore immediately wrote a memo to these people asking for their support in my efforts to reverse the decision, but received no reply from any of them. Not even Art Nozcik, the other Senior Research Fellow, or Dr. David Johnson whom I had recommended as replacement for me as Branch Chief offered to help me. In fact, Dave had visited Marion a few days before the "fait accompli" and for-warned her of my impending dismissal. The reason he gave to

Marion was that "Frank is not a team player." Marion reminded him that I had often pulled the chestnuts out of the fire for people in the branch and Dave replied "He should have let them burn."

I then phoned the Institute director, Dr. Hub Hubbard, and asked for an appointment to discuss my termination. His secretary told me that Dr. Hubbard was extremely busy negotiating the budget for the next year and he would not become involved in the personnel decision to terminate any particular employee. But, the secretary also intimated that if I were willing to retire voluntarily and not challenge the decision of the new manager, I could expect lucrative consulting contracts. The mention of retirement, when in my opinion I was still a vigorous contributor to the field of solar energy and the Institute, raised a suspicion that my age could be a major contributing factor in the decision to terminate me and would give me a legal handle to challenge SERI. I was in a quandary whether I should accept the idea of forced retirement with lucrative consulting for SERI or challenge the decision to terminate me. Two days before the expiration date of filing a complaint for having suffered age discrimination, I filed a suit against SERI. I made that decision with a heavy heart, angry and hurt for having been dismissed for no apparent reason by an organization that I had faithfully served for over a decade.

My decision to sue SERI for age discrimination was based mostly on a feeling of having been treated unfairly and disrespectfully, particularly in the way in which my dismissal was handled. I felt that I had been a good ambassador for solar energy, having published with a former SERI employee, Dr. Ari Rabl, a complete chapter on "Solar Energy" in the McGraw-Hill Handbook of Heat Transfer Applications; written a chapter on "Natural Convection in Active and Passive Solar Systems" with Dr. R.E. Anderson for *Advanced in Heat Transfer, vol. 18* published by Academic Press; and having prepared an international edition of the book *Solar Design: Component Systems and Economics* with my former student, Dr.Jan Kreider and the famous Dutch solar scientist, C.J. Hoogendorn. I had served as Visiting Professor at the University of California campuses in Santa Barbara (1985) and in San Diego (1987), two highly respected universities, given many talks about the promise of solar energy all over the world and assisted NSF and DOE in evaluating proposals in the solar energy field. There was, in my opinion, nobody in the entire Institute who could match the kind of exposure I had provided over the past four years for solar energy, and based on my record, I felt that my forced retirement was unjustified. Ironically, the tenure at SERI of the manager of the thermal program who dismissed me lasted less than a year. He left under pressure just a few months after having administered the *coup de grace* to me.

The outcome of my suit was a settlement. I explained to the judge that I wanted to continue participating in the development of solar energy and that the dismissal from the Institute for whatever reasons would damage my ability to do so. The lawyer representing SERI did not dispute my factual presentation of having made appreciable contributions to solar energy, and simply reiterated the bureaucratic position of SERI that DOE had not provided a charge number for my work at the Institute. While the suit was in progress, the Solar Energy Research Institute was converted into a national laboratory, which is now functioning as the National Renewable Energy Laboratory (NREL). The lawyer for NREL offered a settlement under which I would retire from my position of Senior Research Fellow, but would then be appointed Senior Technical Advisor to NREL. I would at the same time be provided with a contract that would provide me funding for a finite period and allow me to participate in the development of solar energy on my own terms. One of the stipulations of the settlement was that I should not request a desk inside the Laboratory, and conduct all my work offsite. The settlement offer met my main objective of being able to continue work in solar energy, and the status of Senior Technical Advisor was also attractive. Consequently, I accepted the settlement and my suit was dismissed without prejudice to either side.

I had great expectations of continuing productive interactions with NREL, but those expectations were not fulfilled. The Laboratory lived up to its financial commitment by paying for its contractual obligation, but never once called on me for advice after the settlement was signed. All the same, the settlement provided me for a while with financial support that gave me the opportunity of conducting investigations, mostly in the policy arena, on how renewable energy could be integrated into a future national energy strategy. What I did not realize at the time was that my monthly retirement benefits, after having give more than 10 of my most productive years to the organization, would be less than the fee a lawyer charges for a mere 4 hours of consultation.

After leaving NREL I continued serving as the Editor of the Journal of Solar Energy Engineering, and for this work I received an Award of Appreciation from the Solar Energy Division of ASME. It reads: "For valued services in advancing the engineering profession as Editor of the Journal of Solar Energy Engineering." But the term for an ASME Editor is limited to 10 years and I had to give up my editorship in 1989. My farewell editorial describes the status of solar energy and my future hopes for renewable energy.

Frank Kreith

Journal of
Solar Energy
Engineering

Editorial

Ten Years in Review—A Farewell

With this issue the Journal of Solar Energy Engineering celebrates its tenth anniversary. It also marks the end of my tenure as its first editor. These events are good reasons for celebration and reflection.

The past ten years have seen enormous changes in solar engineering technology as well as in the attitude of the federal government towards solar energy. Ten years ago the U.S. was still suffering from the aftermath of the Arab Oil Embargo and the government was determined to take whatever steps were necessary to become energy independent and lay the foundation for a secure long-term energy infrastructure that was to include solar energy as one of its most important primary sources. There was both public and political support to develop all forms of solar energy as rapidly as possible and the President, as well as Congress, were willing to provide moral and financial support. Bumper stickers on cars all over the country proclaimed support for Solar Power and President Carter convened the Presidential Domestic Policy Review of Solar Energy in which I enthusiastically participated. This review concluded that by the year 2000 solar energy could contribute as much as 20 Quads, equivalent to 20 percent of the country's total energy needs.

Entrepreneurs and academic researchers responded to the challenge. Solar industries, as well as academic research centers for solar energy, sprang up all over the country and young people enthusiastically supported their goals. Expectations for early realization of solar energy were high—in retrospect it seems that these expectations were unrealistic and probably contributed to subsequent disappointment. Because sunshine on earth is plentiful and free, many people expected energy from the sun to be both readily available and cheap. To maintain the support of the public and the momentum of developing such a popular energy source, at DOE in Washington solar prophets, rather than solar engineers, were in vogue. But as engineers we know that the development of any new technology takes time and patience. We also know that there is no free lunch, but that capital investment and competition determine success and failure of new concepts.

The ASME Solar Energy Division perceived, in this situation, both an opportunity and a professional obligation. The opportunity was to help an emerging technology make its transition from the laboratory to industry, and the obligation was to provide a peer-reviewed forum for new ideas and developments. With these goals in mind, the Division took the initiative to propose a new journal and asked me to become its editor.

Looking back, I believe that the journal has fulfilled its goals. With the help of the associate editors and the many peer reviewers the journal has built a solid professional reputation and is widely considered the premier publication in its field. It is receiving articles from the foremost authorities and although it was primarily intended as an outlet for U.S. work, its reputation has extended all over the world. I want to take

this opportunity to thank all the associate editors and reviewers who have given freely of their time, and the authors who have shown their confidence by submitting their work to the journal for publication. I also want to thank my wife Marion who has run the journal almost from its inception.

While the journal was conceived in a period of growth for solar energy, political and economic forces have produced conditions in which concerns about the budget and trade deficits have relegated concerns about energy to a low priority. The solar R & D budget has shrunk from $578 million in fiscal year 1980 to $92 in fiscal year 1989, and the projection for fiscal year 1990 is not much higher. But despite lack of support from the federal government, solar power has made some significant advances in the market. It is not cheap but it has become competitive in some areas. According to Dr. San Martin, Deputy Assistant Secretary for Renewable Energy at DOE, wind energy systems can deliver electricity at an annualized cost as low as 8c/kW-hr, while photovoltaic power has dropped to 25c/kW-hr. About 200 megawatts of solar thermal power plants have been installed by LUZ International, Ltd. in California, and according to Dr. D. W. Kearney, its Vice President, the electric power cost has now achieved 8c/kW-hr. These costs are competitive, especially with nuclear power whose cost for plants that came on line in the last two years is about 12c/kW-hr, according to N. Lensen of the Worldwatch Institute. The cost of domestic hot water heating and of energy saved by passive solar architecture is difficult to specify quantitatively because it varies from place to place, but it too is competitive in many parts of the world with fossil or electric heating and cooling.

Given these advances in solar technology with only minimal federal support, one can only speculate what could have been achieved if support had continued at the 1980 level. But with growing concerns about the environmental impacts of fossil fuel combustion and the public reluctance to embrace nuclear power, the future of solar energy is looking up once again. As a group, renewable energy sources have many attractive features, such as a short lead-time in construction, simple design, modularity, and versatility. Renewable technologies cause only a minimal increase in CO_2 and generate no radioactive wastes. Last, but not least, they are indigenous to the U.S. and their deployment could help to eliminate our unfavorable trade balance as well as enhance our national security.

It is generally recognized that the strength of a nation depends today more on its economic health and political resilience than on its stockpile of nuclear weapons. It was energy conservation, not aircraft carriers, that broke OPEC's stranglehold on the industrial world. But this achievement has not solved our energy problem. Energy is the mainstay of an industrial society and the success of our conservation efforts will be short-lived unless we can match it with new ways of energy generation that do not require an increasing supply of fossil fuels, especially foreign oil.

Unfortunately, our success in breaking the Arab Oil Em-

178

bargo that caused a dramatic collapse in oil prices has also weakened our efforts in developing indigenous energy sources and supporting research to develop renewable solar energy. In the past few years, U.S. oil imports have risen again significantly and last year our energy consumption was the highest ever. The 1973 oil embargo and the U.S. reaction to it vividly demonstrated the huge costs and the long time required to change our energy infrastructure. It seems abundantly clear that unless the U.S. develops a long-term energy policy that utilizes all available technologies, another energy crisis is inevitable.

The challenge of energy today is twofold: first, to provide for a transition away from fossil oil and gas while keeping the social and economic costs to a tolerable level; and second, to ensure that an appropriate mix of sustainable energy sources will be available for the long-term future. After the accidents at Three Mile Island and Chernobyl, public fears understandably cloud the future of the nuclear fission-breeder option, although it is available for electric power generation. Fusion power still needs to be demonstrated and, at best, is many decades away from large-scale engineering implementation. But direct conversion of solar energy, although not widely used, is within our grasp for the generation of heat and power as well as for the production of liquid biofuels such as methane.

But the implementation of these direct solar conversion technologies, thermal, wind, photovoltaic, ocean, and biomass, will require sustained federal support which is difficult to come by in a period of enormous budget deficits.

A peer-reviewed professional journal such as JSEE has obviously no place in the political arena. But given the political and economic situation, the journal has an important mission. Although the U.S. tax and energy policies have temporarily halted the transition from R & D to industrial implementation and our fledgling solar industry is barely able to survive in a climate of low oil prices and lack of concern for our future energy supply, there is still a great deal of solar energy research in progress all over the world. A journal such as JSEE can provide an archival depository for the results of solar research and prevent "re-inventing the wheel" when conditions change and solar energy will again receive adequate financial and political support to achieve its full potential.

This is a challenge which I now pass on to the able hands of my friend and successor, Dean Robert L. Reid of Marquette University. I wish him and the journal the best of luck on the road ahead.

Frank Kreith
Boulder, Colorado
November 1989

Since I relinquished the editorship, the journal has grown into the foremost peer reviewed publication in the field of sustainable energy. This growth was the result of the expert editorships of the following individuals: Professor Robert L. Reed, Marquet University, Professor Jane Davidson University of Minnesota and Recipient of the 2012 ASME Frank Kreith Energy award, Professor Dr. Aldo Steinfeld of the Swiss ATH University, and Dr. Gilles Flamant, Director of PROMES-CNRS of France . The scope of the journal has been expanded and its title reads today *Journal of Solar Energy Engineering— Including Wind Energy and Building Energy Conservation*. This growth also reflects the growing importance of renewable energy, both in industry and academia, and bodes well for the future.

CHAPTER 32— LEGISLATIVE FELLOW AT NATIONAL CONFERENCE OF STATE LEGISLATURES (NCSL), 1987– 2001

After my forced retirement from the National Renewable Energy Laboratory in December of 1987, I thought that my professional career was finished. I was not too concerned about my financial future, but I felt that the experience I had gained after many years of being deeply involved in the technical aspects of renewable energy should not be lost. I knew that ASME International had a government relations effort and that the society sponsored advisory positions for state governments. The regional ASME Coordinator at the time, Mr. Bruce Colin, knew of my background and suggested that I apply for the position of Legislative Fellow at the National Conference of State Legislatures (NCSL), whose headquarters were in Denver near the State Capital. I did not know much about this position, but its description sounded interesting. Bruce told me that currently ASME had a Fellow at NCSL, but the man who filled the position was not a good match with what NCSL needed and that NCSL had informed ASME that the organization were thinking of not sponsoring another ASME fellow again. Then he asked me if I would be willing to try and reverse this decision by accepting a one year appointment and during that time meet the expectations of NCSL. He told me that NCSL wanted an experienced engineer to interact with the legislators in all of the 50 state governments and provide them with up-to-date information on technology and the environment. As soon as Bruce told me about NCSL, I realized that my past experience would be an ideal fit with the expectations of the ASME Legislative Fellow. After all, I had worked as an engineer for many years; had written books and articles; had been the co-director of a major NSF program on the interaction between technology, law, and politics; had worked for the Department of Energy, at least indirectly; and had the ability of quickly absorbing information from different sources and integrating it into an objective whole. In fact, I thought that the latter was my strongest trait. Therefore, I accepted this challenge and I started my new position as the Legislative Fellow at NCSL a few weeks later in 1988. The stipend of the fellow was only a fraction of what I might have earned at SERI, but the opportunities and satisfaction I derived from my work at NCSL were enormously greater and more satisfying than money could pay for.

Then need for unbiased technical information for legislators had become a dire national need after congress under the leadership of Newt Gingrich zeroed out the budget for the Office of Technology Assessment which until that time had for many previous years provided information about science and engineering development to members of congress. This source of unbiased information from experts was also indirectly available to state legislators. An indication of this situation is that in 2014 out of 435 members of the U.S. House of Representatives and 100 in the Senate only 11 had engineering degrees. The percentage of members with engineering degrees in state houses is even smaller.

Little did I realize when I began my one year fellowship in 1988 that I would remain at NCSL for the next 13 years and provide information for state legislators on many of the areas that were of interest to me, including, environmental protection, energy conservation, renewable energy technologies, solid waste management, transportation, and electric utility restructuring to all of the 50 state governments. As I learned later, NCSL was expected to respond to telephone calls, e-mails, or any other type of inquiry from virtually every legislator in any one of the 50 state governments. Our response turnaround time was sometimes exceedingly short. But even providing information to a legislator who is confronted with deciding on a bill already on the floor that there is not sufficient information to give a quick answer is sometimes important. For example, one time the year 1990 I received a call asking whether or not the high tension wires in a utility transmission system could cause leukemia or other types of cancer. A possible connection between EMFs and cancer had been hotly debated since 1979 when a statistical study by a Boulder resident showed that children in a school near some high tension electrical wires had a higher incidence of leukemia than average. Some people and realtors in the North Boulder neighborhood asked that the power lines should be buried, a task that would have cost millions of dollars. I told the legislator, who had been asked to pass a bill that could have condemned land close to high voltage transmission lines, that we needed some more time to study this question, but that at this point it would be wise not to take any action. The legislator listened to me and held the bill. After careful examination of the available data, (published in *Science*, Nov. 8, 1996, p. 910) the National Research Council concluded that there was absolutely not a shred of evidence that there existed any causal relation between cancer and high tension transmission lines. The legislator then pulled the bill and probably saved taxpayers an enormous amount of money and grief, because if the bill had passed, it could have ended up in endless legislation.

NCSL also was involved in presenting seminars to various state governments and providing personal testimony in the fields of expertise of the NCSL employees. I was the only engineer in the organization who could provide this field of expertise. The next thirteen years at NCSL were the culmination of all of my professional background and experience and offered me an opportunity to feel that I was "finally making a difference." In addition to providing telephone and e-mail information to inquiries, as well as writing legislative reports, I gave personal testimony in my fields of expertise to legislators in Alaska, Alabama, Arizona, California, Colorado, Delaware, Florida, Iowa, Kansas, Kentucky, Minnesota, Nevada, New Mexico, North Carolina, North Dakota, Ohio, Oklahoma, South Carolina, Tennessee, Utah, and Wyoming, as well as in Washington, D.C. The requests for information included issues in waste management, solar energy, utility restructuring, energy conservation, environmental degradation and transportation.

Soon after starting my new position I realized that I had to learn a lot about communication with legislators. The first time I gave testimony to a legislative committee on renewable energy in Iowa I entered a room with at least 30 people. But as I presented my carefully prepared remarks, one after another in the audience disappeared, until at the end only five people were left. I asked one of the remaining senators with whom I had established a personal relationship "what happened?" He responded "You sound like a professor and these are not your students." I took this remark to heart and thought back to my thespian experience on the Nomad stage. My theater experience must have helped, because from that time on I improved my presentations and made them accessible to legislators and their staff, instead of trying to impress the listeners by showing what I knew about the topic, a common mistake of academicians giving testimony before congressional committees.

In order to ascertain what the current main concerns of the state legislators were, NCSL kept track of how many inquiries in specific fields came to the office. When I arrived at NCSL, it turned out that the majority of inquiries concerned municipal solid waste management. This is an area in which I only had peripheral experience, but I quickly learned what the concerns, options, and requirements of a good solid waste management policy would be. It also became obvious that good waste management would be of great importance in any comprehensive energy plan for the future. I was spurred into action when a ship, named the *Mobro 4000*, loaded with solid waste from an eastern city traveled the high seas like the Flying Dutchman, looking for a port where it could unload its florid and smelly cargo. The story of the ship hit *The New York Times* as well as other newspapers and the unsuccessful efforts to find a port to unload the cargo also became a topic of conversation among state legislators. Although the ship was finally able to get rid of its putrid cargo in one of the small African countries, legislators from many states were determined to prevent a repetition of this kind of situation and asked NCSL to help them.

In response to these requests the director of NCSL asked me to organize a national conference dealing with solid waste management. I complied somewhat reluctantly, but with the help of the competent NCSL staff that had experience in holding such conferences, I was able to organize a two day conference on Integrated

Solid Waste Management in June of 1988 in Breckenridge, Colorado. The conference was attended by representatives from most of the state legislatures and laid the groundwork for a national policy on environmentally acceptable solid waste management. Afterwards I collected the testimonies and presentations from this conference and combined them into a book entitled *Integrated Solid Waste Management* which was published by NCSL. The continued interest and the need for additional information gave me the impetus to expand the original NCSL book into the *Handbook of Municipal Solid Waste Management* which was published by McGraw-Hill, Inc., in 1993. The book was timely and received the 1994 Award of Excellence from the American Association of Publishers.

The philosophy of the handbook was that an integrated waste management approach is not a hierarchical scheme, but synergistic in nature. In other words, an appropriate design of a waste and toxicity reduction or a recycling program, which removes heavy metals such as lead, mercury, and cadmium, as well as batteries from the waste stream, is not merely a reduction and recycling function because it also assists in a possible waste-to-energy incineration function, which benefits from the absence of heavy metals and batteries. Waste reduction and recycling are probably the most positively perceived and doable of all waste management steps because the benefits are so obvious. They save precious finite resources, lessen the need for mining, lower environmental impact, and reduce the amount of energy and material consumed by society.

The book also emphasized that reduction is not a complete process unless the legal and institutional framework creates a market for the recycled products that can beneficially reduce the material picked up from the curb. The technical and engineering functions of waste management cannot function in a vacuum and must be integrated with political and social ramifications of its action. In this respect, NCSL, which had the ear of legislators, was in a unique position to implement a synergistic approach to waste management. Our work in waste management also had a direct impact on the energy field because thermal energy can be obtained by incineration of parts of the waste stream, and it can also be obtained in the form of methane from landfills, which are a necessary final part of any waste management structure.

The third piece of the integrated waste management program—the waste-to-energy combustion—is attractive because it reduces the volume of waste dramatically and can also recover energy in the form of steam or electricity. The major concern of waste-to-energy incineration is the relatively high degree of sophistication needed to operate safely and economically and the fact that the public is leery of its safety. However, my experience in Trondheim, Norway had shown to me that waste-to-energy incineration can be achieved safely and economically, provided the plant is properly designed and heavy metals are removed from the waste stream prior to combustion. In Trondheim, the heat from the waste-to- energy facility was actually used in a citywide heating system, which conducted the steam from the energy combustion facility into the homes in town for heating them. In the near-Arctic environment of Trondheim, there was very little waste of the heat in the summer because heating was required almost 10 months of the year. This approach proved economical because almost all of the heat could be used. In Oslo, a different approach

was taken. The City of Oslo converted the waste into small pellets, which were stored and then could be burned during the winter for heating. These are fairly advanced technologies which are not yet widely used in the United States.

Landfills are the one form of waste management that nobody wants but everybody needs. There is simply no waste management technique that does not require landfilling; however, the technology and operation of a modern landfill can assure protection of health and the environment. Moreover, landfills can also be a resource for methane in a properly designed landfill after it has been filled and closed. But siting a waste incineration facility in the United States has proved difficult. The famous NIMBY syndrome (not in my backyard) is quite prominent in this country. But, I knew from my experience in Trondheim, that it is possible with proper engineering design to have an incineration facility which does not create air pollution. Moreover, removing the methane from landfills is not only useful as an energy source, but also reduces the possibility of an accident because the methane in a landfill can combust spontaneously. Despite the obstacles imposed by public opinion, management of the municipal solid waste stream is no longer a problem. Taking care of waste from industrial facilities, however, still has a long ways to go and probably will require mandates and legal measures to avoid being a source of pollution.

An expanded second edition of the book was published in 2002 under the co-editorship of the famous engineer/scientist from the University of California at Davis, Dr. George Tchobanoglous. He provided not only significant additional technical material, but also a wide array of photographs which he had taken in his long experience as a solid waste management engineer. The book has served as the guide for waste management world-wide and was translated into Chinese. In 2013 one of my Chinese graduate students told me that it serves as a guide for solid waste management policies in that country.

Fairly soon after the success of the national conference on solid waste management, NCSL asked me to increase the scope of my activities. ASME was happy to see that NCSL was expanding its coordinative activities with the ASME Government Relations effort and offered to increase my fellowship stipend. But I told ASME that I would rather see that money be used to establish an internship at NCSL so I could hire an assistant because my workload was increasing enormously. ASME complied with my wish and in the subsequent years I had the pleasure of working with some outstanding intern-engineers. One of these interns, an engineering student from Denver University, Ms. Dena Sue Potestio, returned to NCSL later on and made some important contributions. She is now Vice-president of a University in Kansas.

In the following years, NCSL was able to expand its activities into the fields of solar energy, energy conservation, utility restructuring, alternative fuels and transportation. I could probably fill an entire book about my experiences at NCSL. In a letter which Dr. T. Dwight Connor, the Group Director for Energy Sciences and Natural Resources at NCSL wrote he described work and the background of NCSL as the official representative of the nation's 7,424 state legislators and the only organization in the United States that provides unbiased information and assistance on technical issues to all the state governments and their legislators. The final paragraph of a description of my work reads, "In his work with state governments, Dr. Kreith has been a strong and enthusiastic proponent of science and research, but at the same time he has been a fair and objective analyst of the technologies proposed to deal with major challenges of our day such as environmental pollution, waste management, energy conservation, energy technologies, and viable transportation. His NCSL publications over the past few years testify to the contribution he has made in objectively evaluating science and technology for state legislators and the public."

In 1992, I had an opportunity to describe the role of engineers in public policy decisions. In recognition of my work at NCSL The American Society of Mechanical Engineers awarded me the Ralph Coates Roe Medal " for your significant contributions to a better understanding of the engineer's worth to society through teaching, writing, public lectures, television appearances and provision of technical information to state legislators on waste management, renewable energy and conservation." The medalist is asked to present a lecture as part of the Ralph Coates Roe Award. I had been at NCSL at that time for 3 years and had thought a lot about how information requiring unbiased technical information could be conveyed to political decision makers. Such information is particularly important for a long range sustainable energy policy. I had been a witness to promises such as nuclear power will be too cheap to meter, energy self-sufficiency in 20 years, restructuring of utilities for the benefit of society, and other similar unfulfilled expectations. For the Ralph Coates Roe lecture I tried to summarize my experiences and provide guidance for the future. In order not to break the continuity, the speech has been placed into Appendix I.

My career at NCSL would not have been successful without the active participation and support of my supervisor, Dr. T. Dwight Connor. We seemed to hit it off from the very beginning and our relationship developed from a purely bureaucratic interaction into a personal friendship that has endured. Dr. Connor had earned his doctorate in geography, but his understanding of the basic technical issues related to energy and the environment was sufficient to give me the support needed to provide entry into the halls of the state decision makers. I owe Dwight Connor an enormous debt for the wonderful experience at NCSL that not only gave me professional satisfaction, but also provided me with a feeling of finally having an opportunity to sow the seeds for a rational approach to the topics of my concern and professional attention over the previous 40 years: a sustainable energy system, a rational transportation system that would wean the country from its dependence on imported oil, a healthy environment, and an environmentally acceptable way of disposing the wastes of an industrial society.

In my entire interactions with Dwight, there was only one incident of potential conflict. I had prepared a proposal to the DOE for assisting state legislators in energy planning. Dwight asked me to include a young man by the name of George Burmeister who had just recently joined NCSL as a co-director in the proposal. A few weeks later NCSL heard back from the DOE .It had approved the proposal, but a local DOE manager told NCSL administration that the approval of the proposal was contingent on my being removed from the co-directorship and George Burmeister named the sole director. My first reaction was anger and I told Dwight that I could not accept this arrangement. Dwight then asked me to reconsider because our division needed the money and the activities of the proposal were worthwhile. This was the first and only time in my life that I acceded to a request that was more like a demand. But I could not refrain from trying to find out what caused the DOE to make this highly unusual request and asked for the name and an interview with the local manager of the DOE who made the decision which essentially blackmailed NCSL. I went to see the woman and asked what caused her to issue the ultimatum. She said, "You have a reputation as a 'loose cannon' from your days at SERI where I was a program manager." I asked her what gave me that reputation and she said, "You never listened to our people in Washington, but always followed your own ideas and priorities." When I heard that, I told her "I could not have received higher praise for my work at SERI." Looking back at the goals and accomplishments of the Thermal Energy Branch, I believe that the decisions we made at SERI had stirred our research in directions which, although not always approved by Washington, turned out in retrospect to be the very essence of the country's needs.

After I resigned the directorship, the project was funded and made good progress. First we analyzed some of the problems with federal support of renewable energy. We realized from our interactions with state governments that federal support of any particular technology would not receive wide acceptance because renewable resources vary enormously across the country. For example, a tax rebate for wind energy would be welcomed by states like Kansas, North Dakota and Texas who have large wind recourses but not by states such as Arizona, New Mexico and Nevada. In those states support for solar energy will be welcome because these states receive large amount of solar radiation. Thus, on the federal level it would be much more appropriate to provide incentives that could be tailored by states in accordance with their resources and needs. Once we recognize this problem we provide information about the best resources for each particular state and looked for a champion who would then lead legislation that could benefit individual technologies through the legislature. Our preference in supporting solar energy in the states was to get the legislators to accept a so called renewable portfolio standard (RPS) which required utilities to provide a certain percentage of their total electricity production from renewable sources. By placing the mandate in the future and making it progressively larger, the idea gained considerable acceptance throughout the country and today 36 states have portfolio standards in one form or another. These portfolio standards can be tailored to the sources most available in the state and therefore become more acceptable to the industry there. George and I presented these ideas to many state governments and today more than 36 state legislatures have voted for sustainable energy portfolio standards. The most successful of these RPSs is in California and many other states are following its leadership. The California RPS calls for 25% renewable energy by 2016, increasing to 33% by2020. It also includes landfill gas, municipal solid waste, energy storage and anaerobic digestion among the approved renewable technologies. Unfortunately it does not include incineration of municipal solid waste as

NCSL had proposed.

WORLD RENEWABLE ENERGY NETWORK
(Affiliated to UNESCO)

In early 1995, I received a letter from the World Renewable Energy Network (WREN), an affiliate of UNESCO, stating that WREN had selected me as a Pioneer for the World Renewable Energy Congress, which I had not been able to attend personally. But, the recognition showed me that our work in the states had received worldwide recognition and, as the photo shows, I was given the award in June of 1996, when the Congress met in Denver.

4 February 1995

Dr. F. Kreith
Consulting Engineer
1485 Sierra Drive
Boulder
CO 80302-7846
USA

Dear Dr. Kreith

We selected you as a Pioneer for the World Renewable Energy Congress 1994 in order to reward you publicly and give you the proper certification and status during the Congress.

Since you declined to attend the meeting, we were unable to carry out the ceremony in your presence. Therefore, in 1996 your name is going to be at the top of the list and, hopefully, you will attend the meeting which will be held in Denver, Colorado, from 15-21st June 1996. That way, we can honour you properly. This is an honour bestowed on several people globally who have been chosen for their pioneering work and, of course, your work and your status make you top of the list.

Again, I am sorry if you did not see your name among the list but that will be put right in 1996.

Kind regards.

Yours sincerely

PROFESSOR A.A.M. SAYICH
Congress Chairman
Secretary General of WREN

NOTE WREN ADDRESS

147 Hilmanton   Lower Earley   Reading   RG6 4HN
tel (0) 1734 611364   fax (0) 1734 611365

Towards the end of my tenure at NCSL many legislators believed that the most immediate challenge facing this country was its dependence on oil for the transportation sector. That year, the percentage of oil imported from countries such Saudi Arabia, Iraq, and Iran, who are not exactly friends of the United States, had increased to approximately 50 %. It appeared that this percentage was going to continue increasing, because the prediction of Dr. Hubbert in 1973 that the United States would peak its indigenous oil production by the year 2000 had proved to be correct and oil imports had been continuing to increase ever since. The payments to oil producing countries represented an irreversible transfer of enormous wealth to other countries that could undermine the status of the U.S. as a world leader.

The last official publications I prepared with the assistance of 2 ASME interns was entitled "Ground Transportation for the 21st Century," a major study that was co-published by NCSL and ASME Press in 1999. Ron West and I followed this by a significant publication after I left NCSL, entitled "Legislative and Technical Perspectives for an Advanced Ground Transportation System," (Transportation Quarterly, volume 56, winter 2002, pages 51-73). The introduction to this article reads:

"An efficient and economically viable transportation system is an essential part of a modern industrial society. This is particularly true in the United States where the growth of suburbia requires the average American worker to commute daily a considerable distance between home and work. The situation is exacerbated in many locations by a lack of adequate public transport which requires most commuters to travel by private automobile. The use of single occupancy vehicles not only causes congestion, delays, and air pollution, but also imposes a severe economic penalty on many Americans."

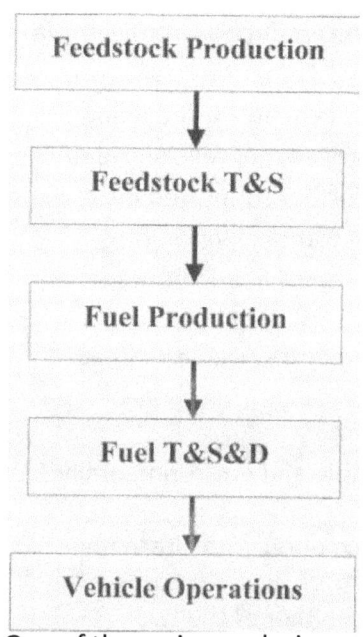

The publication also called attention to the fact that, after housing, transportation is the second largest expense in the average American family budget and introduced the basic idea that any analysis of energy use for transportation should not be based upon the efficiency of the engine, but must utilize what is called a complete "well-to-wheel" vehicle efficiency analysis. This analysis should consist of sequential steps, starting with feedstock mining and/or production, followed by feedstock transportation, storage, and distribution; then fuel production, transportation and storage, finally, the vehicle operation as shown below. Using this approach, we compared ten different technologies with conventional spark ignition engines including the hybrid-electric and hydrogen fuel cell configurations. One of the main conclusions of this study was that the inefficiencies in producing and transporting hydrogen from a central station cannot be offset by efficient fuel cell technology compared to using hybrid-electric vehicles.

Interest and concern about the future of transportation in this country was exacerbated when President George W. Bush, in his first State of the Union address, offered to support a dream of many environmentalists: a pollution-free vehicle that did not need any gasoline. He said, "A simple chemical reaction between hydrogen and oxygen generates energy which can be used to power a car, producing only water, not exhaust fumes," and, he continued, "The first car driven by a child born today could be powered by hydrogen and pollution-free." The idea of using hydrogen to replace fossil fuels was, of course, not new. But the President's pronouncement seemed to the public like a dream come true. But hydrogen does not exist in nature in a form that can be used as a fuel. It occurs in nature only in water (H_2O) or in carbon compounds such as natural gas($NH4$) and must be converted into the combustible form, H_2, by chemical processing. This processing requires energy from a primary source such as coal, natural gas, uranium, wind, or the sun. Consequently, any analysis of the hydrogen vehicle concept needs to take into account the steps necessary to make the hydrogen, then get it into the fuel tank, and finally, to convert it again to electricity to power the wheels. All of these steps waste energy and produce pollution. Hence, to compare a hydrogen vehicle to other options, all of the steps in the technology need to be included. In other words, one needs to perform a well-to-wheel analysis in order to determine whether the President's promises are realistic. Our concern about the policy that the United States was following was further heightened when, on January 9, 2002, the newly-appointed U.S. Secretary of Energy, Senator Abrams, terminated funding for the Partnership for a New Generation of Vehicles (PNGV) Program, on which the Clinton administration had spent a billion dollars in order to produce a hybrid vehicle that could obtain 80 mpg. Despite the promise that the PNGV program had shown, Secretary Abrams stopped further work on this project and instead started a new program called "Freedom Car, "an effort to develop a hydrogen fuel cell vehicle to replace the internal combustion engine. This project was further supported by directives to the U.S. DOE to develop a strategy entitled "A National Vision of America's Transition to a Hydrogen Economy 2030 and beyond." In addition, the National Academy was asked by the Bush Administration to study the path to a hydrogen economy. The National Academy concluded in 2003 that the vision of a hydrogen economy is based upon two expectations:

- Domestic production of hydrogen can be affordable and environmentally benign.
- Hydrogen applications can be competitive in the market with alternatives.

These conclusions seem reasonable enough, but in reality neither of the two expectations has been achieved. Domestic production of hydrogen with any known technology is neither affordable nor environmentally benign and therefore cannot be competitive in the marketplace. Although there was no good technical reason for continuing with efforts to develop a hydrogen transportation system, the President, in addition to his optimistic pronouncement, provided immediately $1.2 billion followed by additional funding from the DOE for hydrogen R&D. As usual, politics backed by money prompted the research community in academia, NREL and other national laboratories to begin preparing proposals to use the available money for research, despite the very serious misgivings that existed among the energy experts throughout the country. The financial success of this "bread and butter" approach to hydrogen research depended largely on the possibility to obtain federal financial support. But in order to find more money for hydrogen R&D funding, DOE had to decrease funding for other energy projects, primarily renewable technologies.

It was, of course, not risky for Ron West and me to challenge the viability of the hydrogen economy proposed by President George W. Bush. Neither of us had a financial stake in the outcome, whereas the financial support of national laboratories and academic institutions depends unfortunately always on the policies of the government. The result of our efforts was a series of articles which, in our opinion, provided an alternate vision for a sustainable and affordable transportation future. The most significant, and probably also most controversial article, was entitled "Fallacy of a Hydrogen Economy: A Critical Analysis of Hydrogen Production and Utilization." [ASME J. Energy Resource Technology, 2004, 126, pp. 249-257.] This article presented a critical analysis of all of the major pathways to produce hydrogen and utilize it as an energy carrier to generate heat or electricity. Once again, it was based upon a cradle-to-grave analysis, including the production of hydrogen, the conversion of hydrogen to heat or electricity, and finally the utilization of that heat or electricity for a useful purpose. This methodology showed that no available hydrogen pathway, irrespective of whether it would use fossil fuel, nuclear fuel, or a renewable technology as the primary energy source to generate electricity or heat, could be as efficient as using the electric power or heat from any of these sources directly. Applying this approach to the transportation problem, we demonstrated that electric vehicles using batteries to store electricity are, in fact, more efficient and less polluting than fuel cell powered vehicles using energy stored in hydrogen produced by electrolysis of water. As a consequence, we concluded that there is no reason to build a hydrogen infrastructure, simply because the overall concept of a hydrogen economy with any currently available technology is flawed. Although, at the time, many of the so-called hydrogen experts who had been recipients of the largesse of the economic stimulus package provided by President George W. Bush to promote hydrogen research disagreed, our conclusions have subsequently been upheld by the results of additional studies. But, the bottom line is that six years later, towards the end of the presidency of George W. Bush, funding for a hydrogen economy has diminished and none of the automakers promises a hydrogen vehicle any longer. Although some day in the future research in fusion or some as yet unknown technology may make a hydrogen technology feasible, in the foreseeable future, funding has dried in the face of the ever-increasing agreement of the technical community with our negative conclusions about a hydrogen economy presented in 2004.

Ron and I followed the realization that the hope to use hydrogen as an economically and technically viable alternative fuel to gasoline was unrealistic with a more hopeful analysis entitled "A Scenario for a Secure Transportation System Based on Fuel from Biomass." In this article, we presented a scenario to meet the future fuel needs of the U.S. ground transportation system that would not require hydrogen, could use existing technology and energy infrastructure, and eventually transition to a liquid biomass-based fuel. This scenario was based upon a combination of the reduction of liquid fuel use by means of plug-in hybrid electric vehicles, followed eventually by generation of ethanol from cellulosic biomass. The approach we proposed in this article combined a demand-side strategy based on energy efficient plug-in hybrid electric vehicles that could charge the batteries from existing and future power generation systems during off-peak hours with alternative liquid fuel produced from cellulosic biomass. We demonstrated that plug-in electric vehicles that could drive for 20 miles on electricity alone could easily triple the gasoline mileage of current SI engine vehicles and reduce the petroleum requirements to less than one-third of that for corresponding SI vehicles currently in use. We also demonstrated that CO_2 generation would be drastically reduced by the proposed scenario. The article, however, ended with a warning that it would take at least ten years for a substantial impact of the transition to an energy efficient automobile fleet to be felt and that it was, therefore, imperative that steps be taken to initiate this transition immediately.

In the years since this article was first presented at an ASME meeting I gave several talks around the country to various audiences describing the strategy that we proposed to achieve a sustainable transportation system. Our ideas were embraced by the presidential candidates from both parties in 2008, as well as by the community of engineers that have examined the various options for a future of ground transportation system. Hybrid electric vehicles such as the Prius have achieved market penetration and both the Japanese and American auto-makers have today hybrid plug-in electric vehicles in their showrooms. It seems to me that when the big three automakers begged for a financial bailout from the U.S. government, this would have been an ideal opportunity to make the bailout dependent upon the American automakers not only switching to more efficient vehicles, but also accelerating the introduction of plug-in hybrid electric vehicles. But the trend towards electric vehicles is continuing and President Obama supported this trend in his 2011 inaugural address with the words "With more research and incentives, we can break our dependence on oil with bio-fuels, and become the first country to have a million electric vehicles on the road by 2015." The total electric vehicle sales for 2013 were at 67,232 according to Climate Progress 2/8/2014 which unfortunately is still a long ways from meeting the president's goal. (http://thinkprogress.org/climate/2013/10/07/2743061/electric-car-sales-up-440-percent/),

Cumulative U.S. Plug-In Vehicle Sales

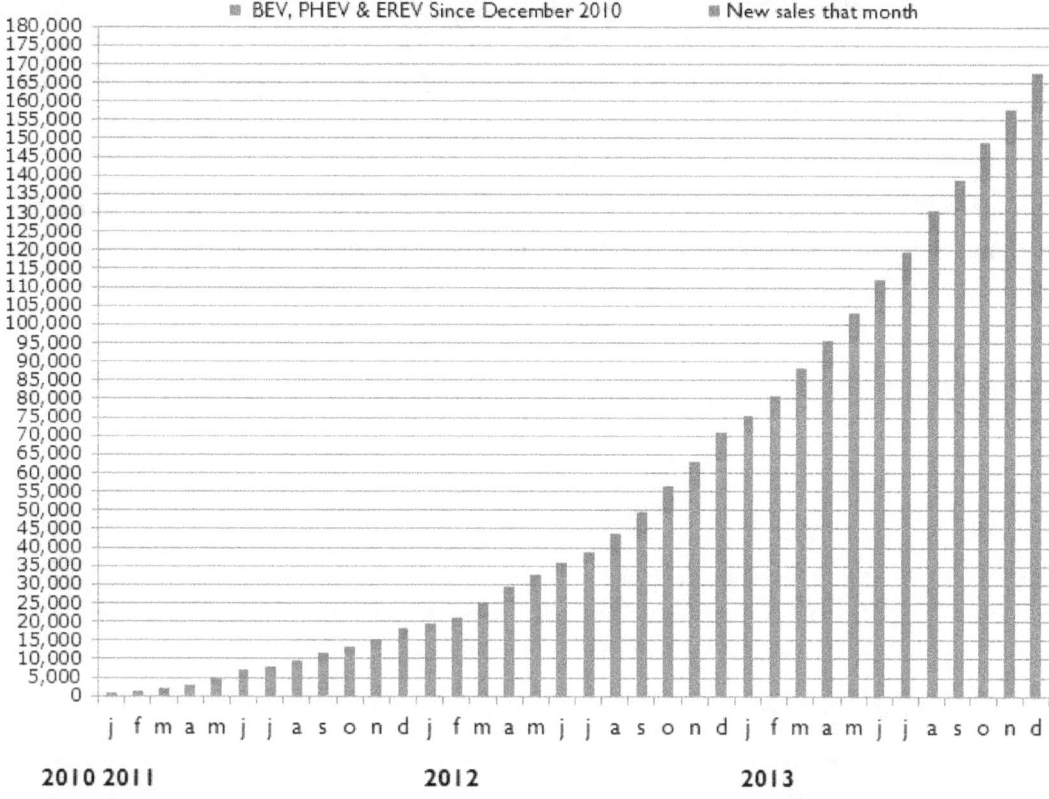

■ BEV, PHEV & EREV Since December 2010   ■ New sales that month

In an effort to make my perspective known to state legislators, the last thing I did at NCSL was to organize a national conference on ground transportation for the 21st century. Although I received approval from NCSL to conduct the conference, I did not get the active support I had received previously from Dwight Connor. The new director, Larry Morandi, and I did not have the close relationship and trust that had existed between Dwight and me. Although NCSL agreed to host the conference, I did

not get a lot of support in raising money, in selecting participants and providing public exposure. The conference was held in 1999. It was fairly well attended, but NCSL insisted on selecting bureaucrats from DC as the key speakers. I was not given an opportunity to select experts that understood the situation and had the know-how of demonstrating to the legislators the crisis in fueling our transportation that faced the country.

I felt frustrated by my lack of control in planning the national transportation conference and decided to use a different approach to publicize the impending transportation crisis. I combined my own notes with those of two interns, DenaSue Potestio and Chad Kimbell, as well as some of the presentations from the conference into a book, which was co-published by NCSL and ASME Press in 1999, entitled *Ground Transportation for the 21st Century*. This book provided an overview of the options and alternatives for a sustainable transportation system, and many of the ideas subsequently have proved correct. These included mandates for increased CAFÉ standards, development of alternative fuels, and the introduction of electric vehicles by the automotive industry. But the problems of sustainable transportation still remain.

NATIONAL CONFERENCE *of* STATE LEGISLATURES

The Forum for America's Ideas

Stephen M. Saland
State Senator
New York
President, NCSL

Ramona Kenady
Chief Clerk, House of Representatives
Oregon
Staff Chair, NCSL

William T. Pound
Executive Director

January 30, 2002

In honor of **Frank Kreith**,

Whose courage in trying to help legislators understand,
Whose patience in watching them struggle,
And whose perseverance in making sure they did not forget,

    Has made a difference.

           From your friends at NCSL.

Denver
560 Broadway, Suite 700
Denver, Colorado 80202
Phone 303.830.2200 Fax 303.863.8003

Washington
444 North Capitol Street, N.W., Suite 515
Washington, D.C. 20001
Phone 202.624.5400 Fax 202.737.1069

Website www.ncsl.org

In early 2001, NCSL decided to move its headquarters from the center of Denver next to the Capitol, a location to which I had been able to commute by public transportation, to a new building that could be rented at a lower cost, but was located on the outskirts of Denver without public transportation access. When I learned about the impending move of NCSL, I submitted on February 20, 2001, my letter of resignation. It was the first time in my life that I had left a major position voluntarily. But I began to realize that my impact had been waning due to a lack of financial support and that with the impending move into the outskirts of Denver, I could no longer expect to commute between my home and the office by public transportation as I had been accustomed to. NCSL gave me a farewell party and a "Good Bye" note signed by all members of the staff. It was a good ending to my career on the state and national energy scene.

CHAPTER 33—REFLECTIONS ON SUCCESSES AND FAILURES AT NCSL, 2002

After leaving NCSL, I asked myself what influence my activities at NCSL had on the development of sustainable energy. Upon reflection, I think that one of the most important roles that NCSL played was the recognition that energy policies could be introduced on a state level much more easily than on the federal level. The reason is that the effectiveness and efficiency of renewable energies technologies varies geographically. For instance, whereas bioconversion from corn is of interest to Iowa, it has no potential in a state such as Arizona which lacks water resources. On the other hand, solar thermal power generation is cost effective in Arizona where there is plenty of sunshine, but not in Maine, where there is relatively little insolation. Similarly, wind energy is of great interest in Kansas, North Dakota and Texas, but not in New York. Hence, once a state recognized its unique opportunities, it was fairly easy to find a champion in the legislature who would introduce appropriate legislation for renewable energy in the home state. Consequently, states became a large-scale laboratory for energy policies because NCSL could evaluate what specific measure worked and what didn't. Since NCSL was charged with assisting all state governments, the organization continuously evaluated the successes and failures of specific policies, and thus I could assist other states by passing along pertinent information and help in drafting appropriate legislation.

My experiences at NCSL demonstrated that it was much easier to convince state legislators to institute energy conservation rather than renewable energy production systems. One of the main reasons is probably that energy conservation measures are fairly simple and are generally useful anywhere in the country, whereas renewable energy production systems are more complex and usually dependent on climate and geography. One of our most wasteful causes of energy loss is the lack of insulation and airtight construction of homes, particularly those built in the aftermath of the Second World War when the veterans returned home. The criterion for success of the building industry at that time was to construct buildings at the lowest possible price, rather than consider the future savings of energy and money.

During one of my presentations in the state of Iowa, I had the good fortune of meeting Senator S. who shared my conviction that energy conservation could be instituted with similar legal measures in all the states. In order to help sell energy conservation measures widely, he agreed to join me in a tour of several states and make presentations on the potential energy and money savings with conservation. Conservation could be sold because one can demonstrate that the financial investment is relatively small and payback for many of the energy saving measures is rather quick. Senator S. had a more folksy approach to sell conservation than I because he could speak the language of his colleagues in other state governments. One perspective I remember that he often espoused was that energy conservation is not more widely implemented because one cannot see energy escaping from leaky buildings. He said to his friends, "If energy were purple and you could see that colorful purple mass coming out of your buildings, you would run back home and put proper insulation in the roof and install energy efficient windows". But energy is invisible and therefore it takes a good deal of effort to convince people that it is worthwhile to conserve it.

NCSL certainly has had a major influence on solid waste management. After the National Conference on Integrated Solid Waste Management held in Breckenridge in June of 1989 under the auspices of the U.S. Environmental Protection Agency (EPA), ASME, and NCSL, I prepared first an NCSL document proposing an integrated approach to municipal solid waste management and then expanded the document into the *Handbook of Solid Waste Management*, the latter with the editorial assistance of Ms.

Bev Weiler. Both the NCSL booklet and the handbook were widely distributed and read by legislators across the country. Slowly, each state began to realize the advantages of an integrated waste management program. Although the concept had

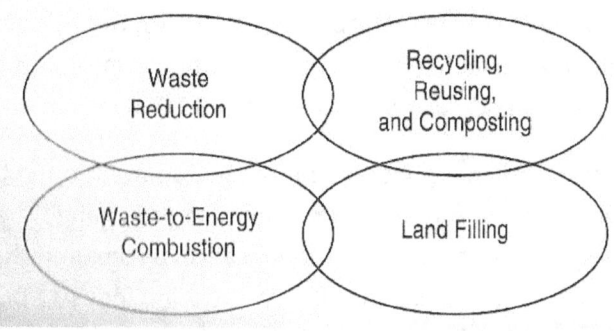

been proposed originally by EPA and was not original with NCSL, the unique position of a state oriented organization made it possible to spread the gospel of integrated waste management into every state of the union. Today every state in the U.S. and other parts of the world as well have adopted, in one form or another, integrated waste management as the means of disposing of its household solid wastes. The motto "Reduce, Recycle, Combust-and then go to a Landfill" has become symbolic.

The popularization of the integrated waste management structure was fairly easy because the same sequence of steps applied to every state. On the other hand, as mentioned previously, recommendations for introducing renewable energy technologies could not be the same for every state. However, the potential of any one specific renewable energy technologies extends over many parts of the country. For example, wind energy can be competitive in several states. Therefore, I organized on April 26th, 1993 as part of the annual ASES/ASME Solar Energy Forum, a roundtable describing the opportunities for wind energy. The roundtable demonstrated that wind energy was coming of age economically. At the time, more than 16,000 wind turbines were installed with an aggregate power rating of 1,500 MW. Although these numbers pale when compared to the enormous growth that has taken place in the last 15 years, they were impressive even then and laid the foundation for other states to introduce legislation to promote wind power. These ideas were clearly explained by two panel members: State House Representatives Patrick Dougherty from Missouri, and David Osterberg from Iowa. Other people serving on the panel were Edgar DeMeo of the Electric Power Research Institute, Jamie Chaplin of the OEM Development Corporation, and Dan Ankona of DOE. An NCSL booklet summarizing the recommendations of the panel was widely distributed among state governments. In addition to the economic opportunities, the booklet pointed out the great potential for wind energy to reduce air pollution, global warming and create jobs in the electric power generation field. Wind energy will create locally four times as many jobs as nuclear power, and probably more than solar photovoltaics or thermal power. Furthermore, the reduction in cost for wind energy projected at the roundtable has materialized and the encouragement that states received, both from the Federal government and from NCSL probably helped in promoting nation wide deployment of wind energy. For example, according to the *Daily Camera* of October 15, 2008, Colorado Governor Bill Ritter said that the growth in renewable energy that his administration has promoted is paying off economically and has generated about 90,000 jobs in Colorado.

Interestingly enough, the conference also mentioned that the most problematic part of a wind turbine is the gear box necessary to change the speeds between the rotor, which is most efficiently between 35 and 50 rpm, and the electric generator that turns at 180 rpm. This is still a problem in the wind industry because these gear boxes have a life of only about five years compared to the anticipated life of 25 to 30 years for the rest of the turbines. All the same, a survey of the installed cost and the capacity factors of wind turbines prepared by Lawrence Berkeley Lab in 2008 demonstrated the enormous growth, reduction in cost and improvements in performance over the past fifteen years, as predicted by NCSL. Capacity factors between 35 % and 45 % are well within range of modern wind turbines and in 2012 the wind power price was below $4 / MWh for a majority of the projects surveyed.

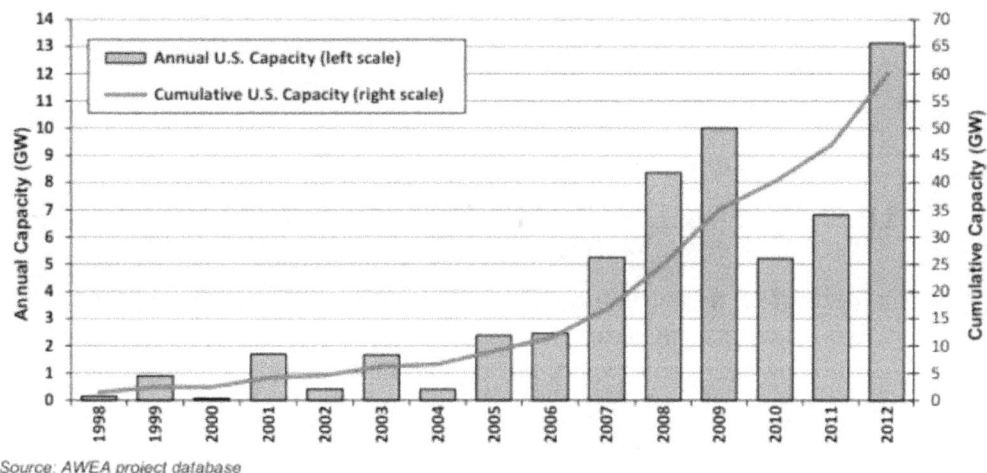

Source: AWEA project database

Figure 1. Annual and Cumulative Growth in U.S. Wind Power Capacity

The impressive deployment of wind turbines is of course mainly attributable to the engineering advances in wind turbine technology, but the confidence of state legislatures in the technology has also been an important factor in this phenomenal growth which is still continuing.[1]

In the field of solar thermal energy, NCSL was able to play a unique role by disseminating information about improvements in the technology and the kinds of incentives enacted by various state governments that had been successful in promoting growth in solar thermal water heating, as well as PV and solar thermal electric power technology. This role was important, because after the reorganization of SERI and the alignment with DOE after its transformation into the National Renewable Energy Laboratory, NREL was no longer permitted by the Reagan Administration to serve an information dissemination function to the public. This function was, therefore, left to other organizations - in particular, to NCSL. In the State Legislative Report of February, 1996 on solar power, NCSL provided important information on the reduction in typical costs of PV systems and the legislative action that states had taken to promote PV. I personally had the chance to explain the opportunities for solar energy in a restructured utility system to professional engineers in my Yellot Award Lecture, presented at Solar 2000 in Madison, Wisconsin. The speech is in Appendix II.

---

[1] Annual Report on U.S. Wind Power Installation, Cost, and Performance Trends, 2012.

CHAPTER 34—RENEWABLE ENERGY IN DEVELOPING COUNTRIES, 1974–2003

Once the population and economy have overshot the physical limits of the Earth, there are only two ways back: involuntary collapse caused by escalating shortages and crisis, or controlled reduction of throughput by deliberate social choice.[2]

In the course of my work at SERI and NCSL I had the opportunity of observing the progress of solar energy in several developing countries. In the following account of my experiences with the development of renewable energy in developing countries, I am deviating from the organization of the book because I believe it will be much easier for the reader to follow events sequentially then by in any one country separately. There were enormous differences between the countries. In many instances, the status of renewable energy was of such elementary nature that in interactions with energy engineers it hardly surfaced. My experiences in these countries demonstrates the enormous difficulties that will be encountered in introducing renewable energy in poor countries where there is hardly enough money to sustain life and where shortages in food and water are often at the top of the agenda for survival.

Yugoslavia, 1974-1985

The first of the countries in which I was able to observe the development of energy sources was Yugoslavia. The first time I went to Yugoslavia in 1974 the country was a federation of six states and two autonomous provinces. The industrial development of the states varied enormously, from the highly industrial northern country of Slovenia, to less developed countries in the south along the Adriatic coast. In Macedonia, water buffaloes were still used to pull the plows in the field. The federation of these countries was held together by the strong arm of General Josip Broz Tito, a Croatian who had been the head of the underground partisan army fighting the Germans during the war. Using the armed forces under his command at the end of the war, he was able to unify the six states into a single country, which although Communist in orientation, did not follow the party line laid down in Moscow by Stalin. The country had once been part of the Austro-Hungarian Empire and many places along the coast still had dual names in German and Croatian. Also some of the

---

[2] Meadows, D. H., et al., "Beyond the Limits - Confronting Global Collapse - Envisioning a Sustainable Future"; Chelsea Green Publishers, Post Mills, Vermont, 1992, pg. 300.

older people were still fluent in German. There were three main prevalent religions among the citizens of Yugoslavia: Catholic, Muslim and Russian Orthodox. Although there were many elements of nationalist pride, intermarriage among all the people of the country was frequent and largely successful. During the war a fascist organization had become very powerful, especially in Croatia, and most of the Jews in the country that were not able to escape were exterminated in the gas chambers of Auschwitz. The educational level in the country was uneven, but the university in Belgrade was generally held in high esteem in Europe.

The International Center for Heat and Mass Transfer was established by a number of the outstanding engineers in the field in order to provide a place for international communication on heat and mass transfer and associated technologies. The founders of the institute decided to hold yearly meetings at one of the most unusual and attractive places in the world. They chose the old city of Dubrovnik, which is located on a peninsula jotting into the Adriatic. When I once came by ship from Greece to Dubrovnik I could see the walls bathed in early sunlight and looking at the city felt as though I were placed into the seventeenth century when the walls were constructed.

Two Yugoslavian engineers were responsible for making all the arrangements that kept the center in operation. One of them was Professor Niam Afgan, a devout Muslim from Zagreb, and the other one was Professor Zarich, a Serb from the University of Belgrade. Both were members of the Boris Kidric Institute of Nuclear Studies in Yugoslavia. Despite their great differences in culture and religion, they believed that the institute would serve a useful purpose and cooperated in keeping it going for many years. I have visited the institute several times and benefitted greatly from this experience.

I met many interesting people from all over the world at the conferences in Dubrovnik. Two of them continued to have an influence on my life and an influence on each other. The first person of importance was Professor Charles D. Hoogendoorn, one of the founding members of the institute. He held the heat transfer chair at Delft University of Technology in the Netherlands and was also interested in solar energy. Professor Hoogendoorn, Jan Kreider and I collaborated on a book entitled Solar Design. The other person that I continued to interact with was Professor Kemal Hanjalic from the University of Sarajevo. We both had an interest in natural convection heat transfer and he invited me to Sarajevo to work with him, and he visited me in Boulder several times. When Sarajevo was successful in attracting the 1984 Winter Olympics, the city wanted a mayor that would have no political association and at the same time would command respect in the community. The city council chose Hanjalic to be the mayor at Sarajevo, a position he held for two years. Afterwards he became the Minister of Science of Bosnia and Herzegovina, one of the countries formed after the Yugoslav federation collapsed. But when the Serbian army encircled Sarajevo and made life precarious in the city, Professor Hoogendoorn invited Professor Hanjalic to join the faculty at Delft in 1994 where he remained until his retirement as professor in 2005.

The meetings in Dubrovnik were quite unusual because it was the only place at the time where engineers from the Soviet Union and the United States could meet and talk to each other. The Soviets did not allow their engineers to travel to countries where they could defect, but Yugoslavia under Marshal Tito was ostensibly a Communist country with an extradition treaty under which defectors would be returned. The United States was supporting Yugoslavia because it did not want it to move into Stalin's orbit. Heat transfer is important for space travel as well as to all kinds of energy production and it was therefore possible to exchange perspectives on energy technology between countries. The Russian engineers coming to Dubrovnik for one of the first meeting proudly told the westerners of their achievements in nuclear power and expressed the opinion that nuclear energy would be a bonanza to uplift the lifestyle of the Soviet people. I had begun to present one of the many short courses on solar energy and the topic attracted the attention both of the Yugoslav and the Russian scientists. Although the Russian engineers seemed to have freedom to talk with others, it was rumored that some of them were being watched by commissars that were keeping an eye not only on the Russians, but also on the foreigners with whom they interacted.

Despite my unfortunate interactions with the FBI which ended in the denial to work on nuclear technology, I remain deeply grateful for all that the United States stands for and has done for me, my family and other Jewish refugees. This gratitude was brought home to me forcefully from a very interesting personal experience I had in the Soviet Union after attending one of the heat transfer conferences in Dubrovnik. I had about a week to spend between the end of the Yugoslav conference and a presentation scheduled in Japan. I consulted with my travel agent as to what might be a worthwhile way to spend this week. She suggested that if I really wanted an interesting experience, different from the usual sightseeing, I should spend the week in the Soviet Union. I thought that her suggestion was intriguing and I asked her to arrange my travel schedule for a week stay in the Soviet Union.

A departure from Dubrovnik by air was always an interesting experience because in the days before suicide bombers, the Yugoslavs had an unusual way of protecting their flights. Each passenger was asked to check in any luggage that he or she intended to take along and before the airplane took off a porter arranged all the pieces of luggage in front of the ramp leading to the plane. Each passenger then had to go and pick up his or her luggage, show the luggage tag to the stewardess standing in front of the ramp, and then take the luggage onto the plane. Those who could not carry their suitcases were helped by an attendant. The assumption of that safety measure was that no one in his right mind would put a bomb into his or her suitcase and then take it on an airplane where he or she was a passenger. This has of course changed and today we have to spend enormous amounts of money on X-ray machines and other devices to avoid having a bomb taken on a plane by a suicide bomber.

After one of the meetings, several of the people were planning to go on to another international heat transfer conference to be held in September in Tokyo. It was therefore natural that we discussed our plans for the interim between the end of the Yugoslav conference and the beginning of the one in Japan. There were several other engineers who were planning to spend the time in the Soviet Union before going on. I asked one of these engineers for the address of the hotel at which the majority planned to stay in Moscow, and wired to have a reservation made for me. However, when I arrived in Moscow and asked for a taxi to be taken to the hotel, I was told that my reservation had been changed to a smaller accommodation which would be equally suitable. I took a taxi to this small hotel where a concierge sitting at the entry way examined my papers and then lead me to a comfortable room upstairs. At the time, I thought that the concierge was merely for convenience of the guests, but soon learned that her job also included scrutinizing not only the people staying at the hotel, but also any persons that may be visiting them. In Dubrovnik I had met Dr. Popiel, an engineer from Poland interested in energy as well as heat transfer. He was with me when my accommodation was changed and suggested that he would like to visit me later that day. We arranged to have dinner together and afterwards go to my hotel which was more conveniently located. When we arrived at the hotel, the concierge asked for Dr. Popiel's passport as well as for his permission to stay in the Soviet Union. After looking over the material, she motioned him to go upstairs and we went to my room. The first thing Dr. Popiel did was to unscrew all the light bulbs in the ceiling and look at all the electrical connections in the wall, while motioning me to stay silent. After he had examined the room for possible listening devices, he relaxed and we spent an interesting evening exchanging our respective experiences about heat transfer, energy and the future.

One of the outstanding attractions of Moscow was the Russian circus. I asked the concierge how one could obtain tickets and she told me that this was very difficult because most of the performances were sold out. I therefore spent the next day in the museums and going on the Russian subway which was quite impressive due to its cleanliness and speed. When traveling on the subway, I was helped by my previous learning of the Greek alphabet, which had some similarities to the Russian script. Consequently, I could identify stations and route my journey to various places of interest.

When I went for breakfast on the third day a young man between the age of 30 and 35 approached me and asked if he could join me at my table. I recognized that he was Russian, but his English was quite good and I was happy to have company. In our conversation, he told me that he had guided some visitors previously and would be happy to show me around. I told him of my desire to see the Russian circus and he said that he would try to get us tickets. In the meantime, we visited a museum and he invited me to his home for dinner. There, I met his family which consisted of a wife and his parents who lived with them in a rather Spartan apartment. We arranged to meet again the following day and he promised to look for a way to buy a circus ticket. His company was very helpful because I could only get a taxi from the front of my hotel and when I tried to hail a taxi on the street I was never successful. But my Russian guide had a way of simply raising his hand, sometimes waving the hat that he continuously wore, and a taxi would stop immediately. Although I paid for the ride, it was certainly helpful to find a taxi in Moscow.

The next day he again invited me to his house, but this time I found a different group of people. There was another couple and an attractive woman who was introduced as a visitor from East Germany. In the course of the evening, my guide took me aside and confidentially said" would you like to spend the evening with this young lady? We have an extra bedroom and this could be arranged". My guide had told me that in addition to helping foreigners, he also had an education in engineering and somehow was aware that I had worked in rocketry in the United States. He only made cursory mention of his interest in rocketry and I did not think much of it. However, when he suggested that I should spend the night with the young lady from East Germany I realized that all his attention may be a trap and I could find myself photographed in embarrassing positions that could lead to blackmail or worse. I therefore told my guide that I thought this was most inappropriate and left immediately. This time I was not able to flag down a taxi, but fortunately my previous experience with the subway came in helpful and I was able to find my way back to the hotel. However, I packed my suitcase and got into a taxi in front of the small hotel and asked to be taken to the large hotel where some of the engineers I had met in Dubrovnik were staying. Although I was told at first that the hotel was full, one of my colleagues offered to have me share his room and I remained at that hotel for the rest of my stay in Moscow. To this day, I am wondering why the Russians should have picked on me. I was later told by a former member of the U.S. Intelligence Service that my experience was typical KGB operating procedures: luring a person with information of potential interest to the Soviets into a compromising position, taking photographs and then blackmailing that person into cooperating with the Russian secret service. For the remaining few days of my stay in Moscow, I kept looking over my shoulder to see if anyone was following me and I did not leave the hotel except with organized tours for foreigners. I was glad when on September 3rd it was time to go to Japan, where I was slated to be the repertoire at the national heat transfer conference in Tokyo.

I was reminded of my brief experience in Russia when I saw many years later (2006) the German drama film entitled "The Lives of Others" (Das Leben der Anderen). This film was released 17 years after the fall of the Berlin wall and gave a realistic picture of what it must have been to live under the constant surveillance of the Stasi, the East German equivalent of the KGB in Russia.

I returned to Dubrovnik several times after my initial visit in 1974. In 1982, I participated in a course on high-temperature heat transfer that was attended by an international audience that included several Russian engineers. Two years later, I was asked by the U.S. State Department to be a goodwill ambassador to the different states that constituted the Yugoslav Federation at that time. Ten years after my first visit to Dubrovnik in 1974, confidence in nuclear power was beginning to wane and the potential of solar energy was widely recognized. With this change of interest in energy sources in mind, the U.S. State Department asked me to undertake a goodwill mission to Yugoslavia and give an hour presentation on the potential of solar energy to the engineering community and then be available to answer questions from the audience for an hour or more. My journey began in the north, in the capital of Slovenia, which in those days was called Ljubljana. Slovenia is the most technologically advanced of the Yugoslav states and at one time was part of Austria. As a result, many of the people there speak German and that made communication easier for me. The following day, a vehicle provided by the State Department drove me and Marion to Zagreb, where I spoke on June 14th and 15th. Then the vehicle went on to Belgrade, where I gave lectures on the 17th and 18th. From there, I went on to Sarajevo, where I gave some lectures and also met up with my friend Hanjalic. After completing my assignment in

Yugoslavia, I rented a car of my own and went on a vacation until I returned to the United States on July 23.

My last visit to Dubrovnik was in August of 1985. At that time, it was already clear that the Yugoslav Federation was in the process of breaking up along ethnic lines. This development resulted in the horrible War of 1992 when the Serbian forces attempted ethnic cleansing of all the parts of Yugoslavia that they wanted to claim for their own country. However, it is gratifying that recently the International Symposium of Heat and Mass transfer has been re-established and takes place yearly in Croatia.

In addition to my professional interaction with Yugoslavia, I had a very interesting personal experience in the country. When the publication of the Principles of Solar Engineering was imminent in 1977, the editor of Hemisphere Publishing, Bill Begell, decided to co-publish the book with McGraw-Hill because the latter had a sales staff that visited universities and could promote the adoption of a textbook. However, there was still the need to provide a cover for the book. Bill Begell suggested that we use the copy of a print by an artist named Ragusin, one of the best known primitive painters in Yugoslavia. Bill was familiar with the primitive painters in that part of the world because his wife had an art gallery in New York. When Marion and I saw the cover, we decided to try and find the original painting for our own home. Our first effort to find the painting was in 1982 during one of the conferences in Dubrovnik, where one of the galleries had other works by the artist. We described the painting and showed the owner a copy of the cover. To our surprise he said that he could obtain the original. However, the Yugoslav government at the time frowned upon exporting its native art and he told us that we would have to surreptitiously meet the person who would sell us the painting at midnight on the outskirts of town. We were told that the art dealer was a reputable man and on his word we met the intermediary and he delivered the rolled up painting upon payment. When we came to our hotel and unrolled the painting, we were unpleasantly surprised. It was indeed a beautiful painting by the artist Rabusin, but not the one that we expected. However, in view of the surreptitious transaction, there was no way of exchanging it. So we kept it.

Our second opportunity to find the painting came in 1984 in connection with a U.S. State Department assignment that took me to Zagreb. There, we did indeed find the very painting we were looking for. Once again, the owner suggested that we not take it directly, but that he would send it with a cousin who was part of the Yugoslav Olympics basketball team who would compete that year. The young man visited us when he came to the U.S. explained that he did not have room to bring the painting because his friends had asked that he bring them several bottles of Slivovitz, a liquor that is unique to Yugoslavia. However, the owner of the painting did manage to send it by carrier and today we are happy to have both of the two originals hanging in our entry hall.

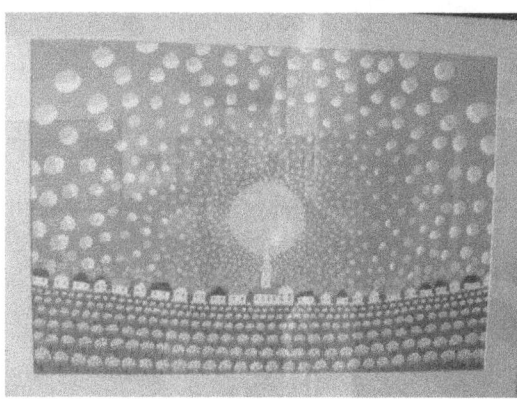

Turkey, 1979

The next country I interacted with was Turkey. Towards the end of 1979 I was contacted by NATO's Scientific Affairs Division and asked to help evaluate a Turkish solar energy proposal. SERI gave me permission to assist NATO and I was appointed a technical advisor to that Division. I traveled to Istanbul, Turkey on November 5, 1979 and visited several industrial companies and the Middle East Technical University. The proposal consisted of three parts, 1- development of flat plate solar collectors, 2- improvement of solar heating and cooling design of buildings and 3- construction of a high temperature solar furnace. In Turkey I found a fairly high level of industrial development, but efforts in solar collector development had stagnated. I interacted with a number of engineers and architects who were well qualified to develop solar collectors. Therefore I recommended that NATO fund the first two of the proposed topics, but not the third since there are already quite a few solar furnaces in operation in other countries. The engineering organization in Istanbul exhibited a good understanding of the potential of solar energy and I believe that the financial support Turkey subsequently received from NATO has been helpful. A report of solar installation worldwide showed that Turkey had by 2012 installed more solar hot water heaters than Germany. But the financial as well as the energy return on an investment in Turkey is considerably better because the solar climate in Germany is much inferior.

I then went on to Ankara where I met with energy bureaucrats to discuss an administrative structure for the program. After our meeting I encountered an interesting question from one of the government official. He said that Turkey admires the United States for its advanced technologies, but wondered whether my advice was an effort to install a second rate technology in Turkey, while the U.S. was expanding construction of nuclear power plants. He asked "WHY, if the U.S. thought that solar energy was so promising, did your country not build solar power plants instead of nuclear reactors?" Quite frankly, I did not have an answer to this question at the time. In retrospect, however, emphasizing solar energy certainly was a good idea and the U.S. stopped building nuclear plants a few years after my visit. Today Turkey has a flourishing domestic solar industry and has not built a nuclear power plant.

India and China, 1992-2003

In 1992, I had a unique opportunity of observing, under vastly different circumstances, the development of renewable energy in what, at that time, were three other developing countries: India, China and Nepal. In the fall of that year, the United Nations invited me to become a consultant in the joint evaluation mission for the Government of India Solar Energy Program. In a background document, the UN mentioned that the development of renewable energy sources had become urgent and essential due to the limited availability of energy from conventional sources. It went on to state that potential for solar energy in India is very good because it is available throughout the year in a large part of the country. Solar thermal energy has therefore been identified as a major contributor to the conventional sources of energy in the future. The evaluation mission consisted of Professor Terry Hollands of the University of Waterloo as team leader and Mr. S.S. Mathur as DOE representative, in addition to me. In order to accomplish our mission, the United Nations provided for a stay of ten days, from January 12th until January 22nd in New Delhi where the Indian Solar Energy Center is located.

After I accepted the appointment to the evaluation mission, I decided that it would be interesting to take a side trip and visit the Hong Kong University of Science and Technology (HKU.S.T), where Professor Tom Stelson had become the Vice Chancellor of R&D. Tom had formerly served as the Assistant Secretary of Energy under President Jimmy Carter and supervised the Solar Energy Research Institute during the 1970s. In addition, I contacted an old friend of mine who had returned to China some time ago and he promised to get me in touch with the renewable energy center in China. I arrived in Hong Kong on January 7th and presented a seminar on solar thermal energy at HKU.S.T, which had opened just a few months previously in Clear Water Bay, Kowloon. In 1992, Hong Kong was still a crown colony of the British Empire, but was slated to revert to Chinese authority a few years later. I was given a tour of the university and was amazed by the high level of research using up-to-date technical equipment, instrumentation, and data collection systems that had been installed in the laboratory.

I also received an interesting briefing on the energy situation in Hong Kong and China where the economy was growing at the phenomenal rate of 12% per year. The limit on growth was not investment capital, but available power. The power situation in Hong Kong was particularly interesting because the people there did not want to build a nuclear power plant in the territory. Instead, arrangements were made to build a large nuclear power plant just outside the territories of Hong Kong in the People's Republic of China. The plant was constructed jointly by Westinghouse and the French government. It provides power now for the needs of Hong Kong, as well as the needs for the rapidly growing industrial center of Guangzhou, formerly Canton.

Upon the recommendation of my friend, Professor Wen-Jeiyang, I had previously contacted Dr. K.M. Guo, the Director of the Guangzhou Institute of Energy Conversion of the Chinese Academy of Sciences, who invited me to visit the institute. In a letter dated December 2, 1992, he informed me that the institute had been working for over ten years on renewable energy including solar, ocean, geothermal and biomass as well as conservation technologies and heat storage. On January 10th, I went by train from Hong Kong to Guangzhou, where I was met by Professor Cheneni-jain, the Deputy Director of the institute. He gave me a guided tour of the facilities of the institute and also showed me some of the products currently under development. The approach of the institute was very different from the R&D of Hong Kong. The institute directed its efforts toward immediately useful renewable energy applications, such as a small solar refrigerator that could be used for storing medicine in isolated areas and for developing simple domestic hot water systems that could be produced anywhere in the country. I was also told that there had been a change in philosophy towards research institute in the PRC. Whereas previously these institutes were funded by the central government, it had become a requirement that the institutes obtain also funding from the local industry and work closely with industry to develop useful products that also could be exported. The facilities of the Chinese institute however were antiquated and Spartan compared to those at the HKU.S.T. There was a scarcity of copper pipe and most of the production machinery, such as lathes and milling machines, were quite old. Nevertheless, the morale of the institute personnel was high and when I presented a lecture on Integrated Energy Management on January 4th to the Chinese Academy of Sciences, it was apparent that although my lecture was being translated, most of the people in the audience seemed to speak English and asked questions in English without translation. The setting for my lecture was Spartan, but the overhead equipment was excellent in contrast to India, where the setting was palatial, but the overhead equipment would not work properly.

From my hotel window, I could see that the transportation at that time was almost entirely by bicycle and I thought that this may be a blessing for our environment because I could not imagine how much air pollution 1.2 billion Chinese would produce if they had automobiles for transportation. This situation has of course changed in the intervening years. Today many Chinese own an automobile or a put-put and the resulting air pollution has become a national hazard.

I also had some interesting discussions with some of the Chinese officials regarding the transition from communism to a form of capitalism in the U.S.S.R. These people felt that the transition in Russia had been too rapid and unstructured, resulting in some of the Russian officials acquiring enormous wealth, especially in the production of raw materials from natural resources, without benefiting the people. They wanted a more gradual transition for China which has taken place in the meantime, albeit with a lot of accusations from the West that there is no freedom of expression in the country.

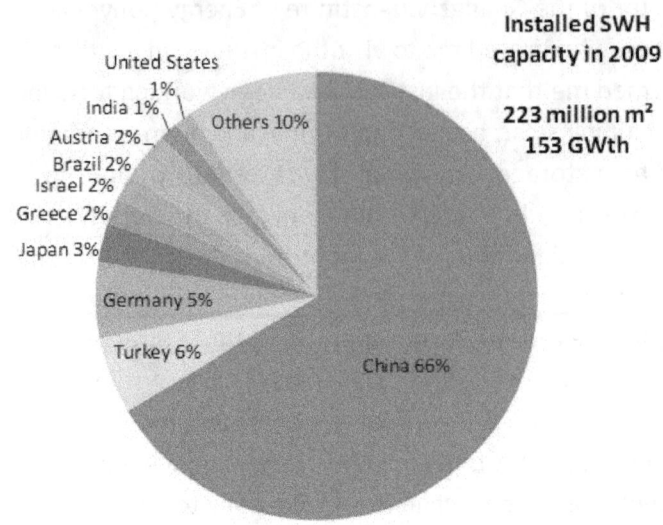

Installed SWH capacity in 2009

223 million m²
153 GWth

The approach taken by the Chinese at the time of my visit seemed more like a dream than realism. However, by the year 2009, China had the largest number of the solar hot water heaters installed and was producing more solar domestic hot water heaters than the rest of the world combined.

China has made equally good progress in the field of photovoltaics. By 2010 China and Taiwan produced about sixty percent of the world's photovoltaic panels (*Daily Camera*, August 15, 2013) and by the year 2013 Mainland China alone produced more than two thirds of all PV panels worldwide. This gave the country a virtual monopoly because of their lower costs, achieved largely with the help of government subsidies. As shown in the figure below, the rise in Chinese PV production was almost exponential. On the positive side it illustrates how quickly an industrial country can rev up the production of renewable technologies if the political decision makers decide to do so.

On the negative side, the growth of the solar industry in the U.S. around 2011 depended to a large extent on the importation of inexpensive Chinese PV panels which were installed and maintained in the U.S. by several small industries. However, the remaining PV manufacturers in the U.S. were afraid that they might go out of business if the inexpensive Chinese PV panels would continue to be sold here and they persuaded Congress to impose tariffs and other trade barriers on the import of these panels. In retaliation for these punitive measures the Chinese government imposed tariffs on American poly-silicon, an ingredient required for the manufacture of the panels. U.S. poly-silicon markets accounted for about a quarter of the total and most of the production was shipped to China. When China imposed a tariff on the poly-silicon from the U.S., prices of solar panels soared and the growth of the U.S. solar industry, which relied mostly on domestic installation and maintenance, went into a tailspin. These developments, as well as the depression in Europe which caused the EU government to reduce buying photovoltaic panels, also created chaos with the Chinese manufacturing sector and Suntec, the largest Chinese PV manufacturer, declared bankruptcy in 2013. This should serve as a cautionary tale for future expansion of renewable technology: make sure there is a market to buy the product!

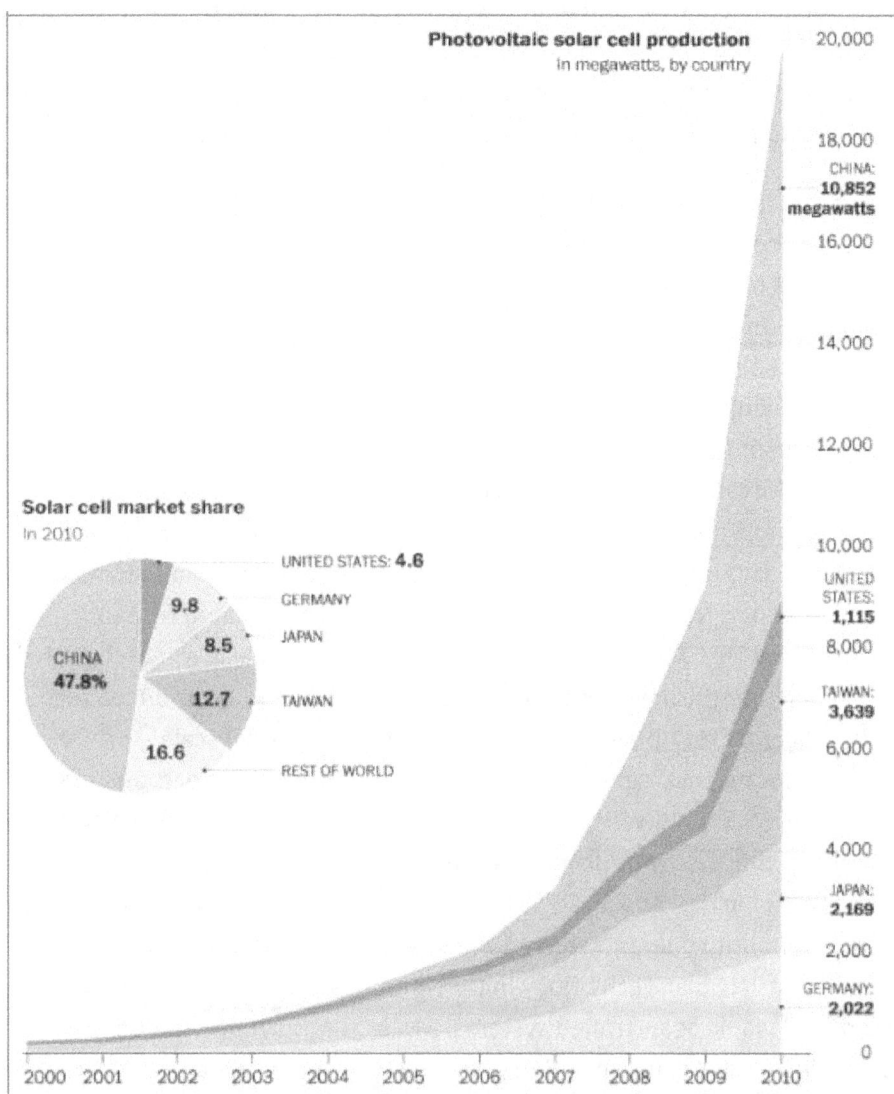

Three days after my visit to China I traveled to New Delhi where I began my assignment for the United Nations. India had made an effort in the development of solar energy. Under the previous Prime Minister, Indira Ghandi, a secretariat for non-conventional energy sources was established and the secretariat was later elevated to a ministry that reports directly to the prime minister. A building compound to house a solar energy research center had been built near New Delhi and an outside development program had provided part of the funds to train the staff, set up facilities, build up a library and buy scientific equipment. It was our job to evaluate the effectiveness of the UNDP contribution to the Indian solar energy program.

India has no fossil fuel resources to speak of and is therefore one of the prime candidates for new energy technologies. India had been successful in setting up wind farms and was beginning (with government assistance) to sell thousands of solar cookers which were effective in reducing deforestations. They had also built many flat-plat collectors for domestic hot water heating with government subsidies and were now planning to move to the area of industrial process heat. However, a large part of the Indian effort was based on basic research which was not directly related to near-term industrial potential. I have also learned that the solar cooker effort has been abandoned.

Our inspection of the way the United Nations' funds had been used revealed an interesting situation. A large portion of the United Nations' contribution had been used to purchase a solar thermal power plant from a German company by the name of MANN. The plant had been built on the Luz principle, which was used in California for the construction of a highly successful 360 MW parabolic trough solar plant, but it had been built in Germany and set up in New Delhi as a "turnkey" installation sometime before our arrival. Unfortunately, the German company went bankrupt after having installed the plant in India and there was no one available to teach the Indian staffs how to operate the facility which was quite sophisticated. Hence, the plant stood idle. About a third of the United Nation funds were used to buy sophisticated research equipment, such as a high-power microscope, which was used in basic research at the center. The remaining one-third was used to help develop indigenous products, such as solar cookers, collectors for domestic hot water heating and small-scale photovoltaics. The last one-third seemed to have produced positive effects and we were introduced to Mr. S.S. Ahluwalia, the managing director of Surya Jyoti Devices India, Ltd., a company that was producing solar energy conversion systems. This company was interested in developing solar trough technology which had reached the point of industrial competitiveness. I referred the director to my former student, Ken May, the CEO of Industrial Solar Technology in Denver, CO at the time, a company that had already produced viable solar trough systems. Mr. May is currently the Chief Engineer of the Spanish solar company, Abengoa, which has built the largest solar thermal power plant in the world. I do not know what Indian officials decided to do after receiving our evaluation report, but India opted for developing nuclear energy as its main power source.

Although I had no further personal interaction with the Indian solar program, I did receive an update on India when Yogi Goswami and I prepared a second edition of our "Handbook of Energy Efficiency and Renewable Energy" published by CRC Press in 2014. In chapter 2.5 of that handbook, Depak Gupta and P.C. Maytaus provided an update on the Indian solar program. It turned out that the Indian government reversed course on solar energy. In 2010, India launched an ambitious program of achieving 20 GW of solar grid power by 2022, 2 GW of off-grid solar power, and installing 20 million square feet of solar thermal collectors. According to this update, the program is under way and by competitive bidding in the world market as well as local initiatives, India has become the lowest cost solar electric power producer in the world.

From India, I went on to Nepal, where I met up with my former student, Mr. G.R. Shakya, and Mr. T.P. Gaushan, the president of the Center for Renewable Energy in Nepal. There I learned that, realistically, the industrial capacity of the country was insufficient to build their own solar equipment and the emphasis was largely on obtaining funds to buy and then install small photovoltaic systems for lighting in schools and powering small refrigerators to store medicine in isolated regions.

My observations from visiting these three developing countries were that there exists enormous potential and interest in renewable energy. However, the old idiom that "give a man a fish and you feed him for a day; teach a man to fish and he is fed for life" certainly applies. In order to see renewable energy grow in the developing countries, the technologies should be consistent with the level at which that country can handle them and then move forward, while starting with highly sophisticated equipment is not productive. I believe that my observations at the time have now been born out and developing countries are developing their own renewable energy programs. The United States was held in high regard by all of these countries and the engineers and scientists closely followed the development, both here and in Europe to guide their progress. Library facilities were quite good, although access to current scientific and engineering journals was limited.

Cuba, 1997–2002.

In June of 1997 I received a letter with an unusually beautiful stamp. Looking at the stamp, I was surprised to read the word "Cuba" and even more surprised on noting that the sender was Fidel Delgado from the Technical Information Center in Havana.

Dr. Frank Kreith, P.E.
Consulting Engineer
1485 Sierra Drive
Boulder, CO 80302-7846
Phone 303 443 1406   Fax 303 443 4341
e-mail FKreith@aol.com

Wednesday, November 17, 1999

Mr. Fidel Delgado
Tech. Information Center
P.O. Box 60
10100 La Habana
Cuba

Dear Mr. Delgado:

In June of 1997 you wrote to my husband to ask for a reprint of an article on renewable energy which he sent to you, and which we hope you received.

I am writing to you at this point because there is a very good possibility that we will be coming to Havana in January - in connection with a course on Cuban dance and rhythms that my daughter has registered for and has invited us to come with her - and I was wondering whether you might be interested in meeting Dr. Kreith for some possible further discussions on renewable energy or some other topic of interest to you. If we go, we will be in Havana for a week, starting January 9th.

I hope that this letter reaches you in time so that if you would like to meet with us, you have plenty of time to get back with us. E-mailing would be the best way to communicate, and the e-mail address is at the top of the letter.

With best regards,

The letter said, "Please send me a copy of your article entitled "Incentives for Renewable Energy Generation in the United States." The article had been published in *Renewable Energy* in September of 1996 and summarized the experiences of NCSL with various means and incentives to stimulate the use of renewable energy. I was amazed that a country, which in the minds of many people was a Communist dictatorship, should be interested in incentives. However, I sent Mr. Delgado a copy of this article with the cover letter shown above, but I did not think much about it until two years later, when my daughter Judy told Marion and me that she planned to go to Cuba for a workshop on salsa dancing and South American rhythms, topics which she taught in the United States. She invited Marion and me to come with her because she was curious to learn more about Marion's experience during the many years she spent in Cuba as a refugee during and after the War. We accepted Judy's invitation with appreciation for her interest in our past and anticipation to revisit Marion's former home. In view of the embargo, which the United States had placed upon Cuba after the Cuban Missile Crisis of 1961, we had to travel illegally. This required first going to Cancun in Mexico and then taking a Cuban airliner to Havana. Upon arriving in Cuba the custom official place a sheet of paper in our p[passport and put the official stamp on that paper which was removed when we departed. Thus our passports did not have any evidence of our trip.

Marion had spent many years in Cuba, where a group of refugees from Belgium had established a diamond industry with the blessing of the Cuban government. As a condition of the permission to build this industry in Cuba, the government demanded that half of the people employed should be Cubans, but the other half could be refugees. In view of the fact that hardly any country was willing to allow Jewish refugees from Nazi Germany to enter, the possibility of going to Cuba was a lifesaver for many of them and the fact that they could earn a legitimate living was a additional bonus. Virtually the entire industry left after the war ended and Castro overthrew the old dictatorship.

Cuba had, at one time, been intended to become a showpiece for Socialism and the Soviet Union had provided financial support to the country by buying the sugar cane it produced at prices well above the world market. However, the support of the Soviet Union came to an abrupt halt when the economic situation of the Russian Federation crashed, and suddenly Cuba was faced with the need to become self-supporting. While the Soviets were providing support, Cuba enjoyed the availability of cheap oil. But once the Soviet aid stopped, Cuba had to set up a new economy with hardly any oil. The transition period was hard and is now called "The Special Period"

Once I knew that we were going to Cuba, I began to read up on its history and realized that, in a way, the experience of the Cubans to become a self-sustaining entity could be a model for other countries when the time comes to drastically reduce the consumption of oil and gas. When I arrived in Cuba, I was introduced to a number of people that told me more about the Special Period. Faced with the inability to buy oil and food from outside, the Cuban government adopted some inventive strategies to survive. The Cuban dictator, Fidel Castro, immediately ordered a million bicycles from China and also provided incentives for all the Cubans to grow food in their backyards and, in some instances, on their roofs. In the countryside, farmers reverted back to using horses and draft animals to pull the plows, a technology which in more developed countries had been totally supplanted by diesel engines and tractors.

Cuba did not fit the mold of a typical developing country. Although the country certainly was poor, it had a good education system available to all and also socialized medicine with well-trained doctors from Cuban universities. In fact, it was well known that the top Russian bureaucrat went to Cuba for advanced medical treatment and the Cubans traded their well-trained doctors for oil with countries such as Venezuela. The Cubans were also very interested in renewable energy and ISES, the International Society for Solar Energy, had an active chapter there. At one point, Fidel Castro decided that he wanted every hospital, even in the outlying mountains of Cuba, to have the means of storing medicine in a refrigerator. Cuban engineers converted an abandoned Russian electronics factory to produce primitive photovoltaic cells and were able to provide solar power to the hospitals that were not on an electric grid. Although many of the Cubans complained, I was told that the overall health actually improved after most of the Cubans were forced to walk or use the bicycle for their main transportation. Cuban chemical engineers also developed a means of using the bagass, which is the waste from the sugar cane, to co-burn with coal or oil in their electric power system. All of these conservation measures made it possible for Cuba to survive virtually without oil and, to many environmentalists, Cuba became a showcase of sustainability. An organization made a movie of this remarkable transition entitled " The Power of Community – How Cuba Survived Peak Oil" I have used that movie as a case study for transition to sustainability in my class on renewable energy.

In the course of our visits to Cuba, we met some remarkable people. One of them was Dr. Elena Vigil, who was the person in charge of developing renewable energy in Cuba. Elena had obtained a PhD in physics in the United States and returned to her home when Castro took over the country. At the same time, all of the rest of her family, who were quite wealthy, fled to Florida and Elena had not seen her family since. I was invited to give some lectures and established contact with the solar engineers in Cuba. After returning to Colorado, I was able to interest a number of solar organizations, including the National Renewable Energy Laboratory, to invite Elena to give some lectures on her specialty, which was advanced material for photovoltaic cells. When Elena came to the United States, she was well received and demonstrated a high level of competence in the technical status of her country. About a year after Elena's visit I received a letter from her. She asked" Would you like to give some lectures on solar thermal energy at an international symposium which is to be held in Havana in 2001?" I emailed her my acceptance of the invitation immediately. In view of the special invitation; we were this time allowed to travel directly from Florida to Havana, a trip that took less than an hour, compared to the whole day necessary to go to Cuba via another country. My lectures were well received and we are still in contact with Elena and her colleagues.

At this time (2014), the embargo is still in place, but the situation in Cuba has changed considerably. Many European travel organizations have built beautiful hotels along the Cuban coast and the emerging tourism is bringing hard currency to the country. Moreover, Cuba discovered oil in its territorial waters and is now able to provide liquid fuel to its transportation and industry. This is in some ways unfortunate, because once oil is available to a country, the people forget that this is only a temporary situation and no longer put sufficient emphasis on working towards sustainability. Nevertheless, the experience in Cuba demonstrates that a transition is possible and we can learn a good deal from its success. Certainly, the standard of living decreased and the food sources had to be restricted to what could be grown domestically. Also, many of the luxuries of the Western world were and still are unavailable to Cubans. Although some of the restrictions have been lifted, Cuba is still not a free country and travel restrictions, especially for well-educated Cuban people, are still in place. But some of the measures imposed after the Russians withdrew their support are indicative of what can be done when a country is forced to find a way to do without massive use of fossil fuels.

We were in Havana on September 11, 2001 when Al-Qaida terrorists struck the World Trade Center in New York. To our amazement Fidel Castro interrupted a speech on education and offered medical help to the victims of the terrorist attack, saying that Cuba knew what it was like to be attacked by terrorists. I think he referred to the repeated efforts of the U.S. intelligence Agency to assassinate him. But the Cuban people genuinely felt sorrow about the loss of life from this terrorist attack. We were scheduled to leave the day after the attack and when we arrive at the airport it was crowded, but there was no panic. We did not know how extensive the loss of life and physical damage inflicted by the Al-Qaida fanatic terrorists was until we landed in Miami.

CHAPTER 35—PERSPECTIVE ON U.S. ENERGY POLICY, 1992–2014

"The American way of life is not negotiable." – George H.W. Bush, 1992, Earth Summit

The use of wood and wind for power began in antiquity and was followed by the age of coal which started in Great Britain in the 18th century. The petroleum era began when the first oil well was drilled in Pennsylvania in 1859, followed in 1949 by discovery of Saudi Arabia's monster oil and natural gas fields. The potential of renewable energy production with solar photovoltaics cells began with Albert Einstein's Nobel Prize-winning publication on the photovoltaic effect in1906 and research under the direction of Dr. Paul Rappaport at the RCA labs in the 1950s. The advent of nuclear power began with the development of the atomic bomb at Los Alamos and the construction of the first nuclear power plant in Russia in 1956. The changes in global energy use are shown below.

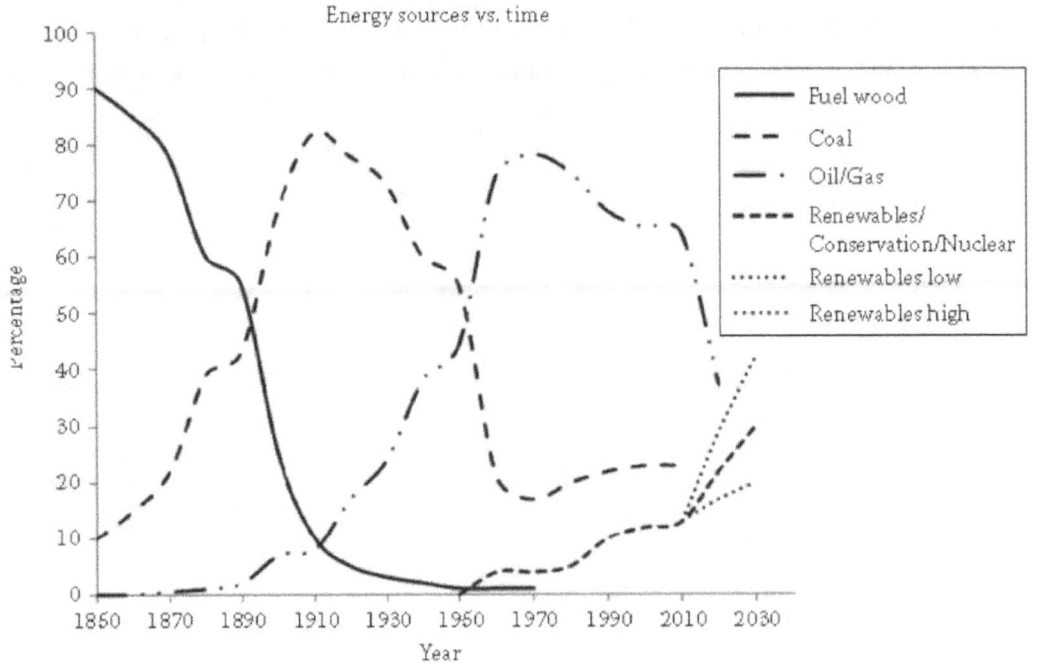

The future scenario is still not clear, but one thing is certain: the order of magnitude of our gluttonous energy consumption is unsustainable and without a major policy initiative the band aid type of solutions that have heretofore been tried will not prevent the impending energy crisis. I am reminded of the situation described at the start of the book when the mindset after the Munich Accord created the illusion of future world peace and ignored the long-time need to prevent world war. Today the small actions taken by decision makers create the illusion of future energy security and ignore the need to prepare for major initiatives to address the long haul.

Before outlining a path for a sustainable energy future, let us step back and look at what the energy policy of previous presidents has been. It is the prerogative of every new administration to develop new policies and a national energy policy is a case in point. The first president to have an energy policy is probably Franklin Delano Roosevelt, under whose presidency the great hydro-dams of our country were built. They laid the foundation for the sustainable electrification of the rural U.S. This is the kind of commitment necessary to carry out a transition from fossil to renewable energy. These dams were immensely expensive in the financial picture of their day. For example, the Grand Coulee dam, which was completed in 1942, if priced in terms of its equivalent share of the Gross Domestic Product of today, would cost somewhere around 40 billion dollars. These dams utilized the most effective locations available for hydro power and most of them are still producing renewable energy today. But since these dams used up the best locations for hydroelectric power, this technology will only be marginally useful for a renewable energy future in the U.S. and other technologies will be required for electric power generation.

In 1951, President Harry S. Truman created a Blue Ribbon Commission to evaluate and propose a plan for the U.S. energy future. The commission, named after its chair William S. Paley, published in 1952 the Paley Commission Report. It proposed that the U.S. energy future should be built on solar energy sources. Surprisingly, the report made an ambivalent assessment for the future of nuclear energy and recommended instead "aggressive research in the whole field of solar energy". The Paley Commission specifically stated that the chief problems in developing nuclear energy for power are:

1) To develop a breeder' reactor that produces fissional material at a rate at least equal to its consumption and

2) To produce an efficient reactor at low cost.

Looking back at the conditions for successful implementation of nuclear energy outlined 62 years ago, it is apparent that neither of the two conditions has been achieved. Prototypes of breeder reactors have been built in the U.S., France and Russia, but none of them have been pursued to a commercial scale for technical and economic reasons. As far as cost is concerned, the cost of nuclear power plants has increased dramatically over time as more and more safety requirements have been mandated. Consequently, the conditions posited by the Commission for successful implementation of nuclear power have not been met. The report also supported R&D for wind and biomass utilization in a national energy policy. (Reference: Blogging about the Unthinkable: What Did the Policy Commission Report Actually Say About Nuclear Power?)

President Eisenhower, unfortunately, did not follow the recommendations of the Paley report, but rather listened to Dr. Lewis Strauss, the chair of the Atomic Energy Commission, who claimed that nuclear power would be the miracle of the future and would be "too cheap to meter." The engineering community at that time had a lot of faith in the future of nuclear power and President Dwight Eisenhower launched the age of commercial nuclear energy with his December 8, 1953 "Atoms for Peace" speech to the United Nations General Assembly. On August 30, 1954, upon signing the Atomic Energy Act of 1954, the president went further and said "under the act our technicians can assist friendly nations in building reactors for research and power." Included among those friendly nations whom we helped to build nuclear reactors were Iran and Pakistan. How times have changed!

Today, despite huge investments and subsidies for nuclear energy [estimated at one trillion U.S. dollars from 1950 to 2007] the future of nuclear energy is bleak because construction of nuclear power plants has turned out to be hugely expensive, there is no depository for nuclear waste and competing renewable technologies have made enormous progress despite the minuscule financial support these technologies have received [estimated at $40 billion from 1972 to 2007]. Even though some politicians talk about a "nuclear renaissance," the fact is that solar and wind technologies have had a faster rate of growth, as well as higher cost effectiveness and increased job creation, than nuclear power over the past few years. Moreover, nuclear accidents in many parts of the world have turned public opinion against building more nuclear power plants.

Richard Nixon, who was elected President in 1970, was probably the first president to publicly announce a national energy policy. Faced with an energy crisis after the oil embargo by the Arab states in 1973, Nixon called for reduced dependence on oil imports and launched Project Independence "to make the U.S. energy independent [of oil imports] by 1980." Unfortunately, the dependence of the American public on the automobile made that effort impossible and in 2013 oil import was twice as much as what it was before the Arab embargo. Neither President John F. Kennedy nor President Lyndon B. Johnson appear to have spent time on an energy policy, probably because they were preoccupied with preventing a nuclear war, dealing with the aftermath of the Bay of Pigs defeat by Cuba, the race to the moon, the Cold War, and finally the Vietnam War.

Most of the U.S. Presidents have been concerned with energy production and conservation as shown by some of their pronouncements below:

- 1973 President Richard Nixon: "By 1980, we will be self sufficient and not need foreign energy."
- 1978 President Jimmy Carter: "In 20 years, 20% of our energy will be from solar."
- 1980 President Ronald Reagan: "Alaska has a greater oil resource than Saudi Arabia."
- 1991 President George H. W. Bush: "Three principals guide our [national energy] policy: reducing our dependence on foreign oil, protecting our environment, and promoting economic growth."
- 1994 President Bill Clinton: "… The nation's growing reliance on imports of crude oil and refined petroleum products threaten the nation's security…"
- 2003 President George W. Bush: "The first car driven by a child born today could be powered by hydrogen."
- 2009 President Barack Obama: "It will be the policy of my administration to reverse our dependence on foreign oil – this time the effort is for real."

Unfortunately, the differences in perspective by our presidents have made it virtually impossible to frame a long term energy policy. It is an undeniable historic fact that no president has left office with the country mining less fossil energy than when he was inaugurated. Since the first energy crisis under President Nixon in 1973, the world has increased its energy use equal to that of another United States on the planet. The staggering scale of the world's energy use is one of the key challenges of our era and it is undeniable that the energy options available to any president or nation are quite limited. But the advances in fracking and horizontal drilling for natural gas and oil since 2010 have made available a virtually new energy source of substantial magnitude that could be used as a transition fuel to build a renewable energy system with minimal environmental costs.

As mentioned before, in the period between 1950 and 2007, about 1 trillion dollars were spent in support of nuclear energy in the United States compared with support for renewable energy of only 40 million dollars in that period. Nevertheless, by 1996 the last U.S. nuclear power plant had been built, partly because of the public's fear of nuclear radiations from accidents. But the main reason for the demise of nuclear power was that the construction of nuclear power plants turned out to be immensely expensive and utilities were unwilling to assume the financial risk for constructing new ones. The increase in cost for constructing nuclear power plants in France, a country that has put virtually all his energy eggs into the nuclear basket, has been quite similar to that in the U.S. Although, the U.S. Congress under the Anderson Act promised to pay for and protect utilities from suits filed in case of the failure of a nuclear power plant, Wall Street is still not willing to provide loans for new ones.

So far all nuclear power plants in the U.S. are operating on a Rankin cycle, similar to a coal fired power plant, except that the heat is generated by a nuclear fission reaction. However, for safety reason the upper temperature of nuclear plants is lower than that for fossil power plants. Therefore, their thermodynamic efficiency is lower and the nuclear plants require more cooling water per unit power output. There are many ideas for different and more efficient ways of nuclear power generation floating among pro-nuclear engineers, but none is anywhere near commercialization. Nuclear energy may someday play a part in the transition to sustainable energy, but it must first overcome the following engineering challenges:

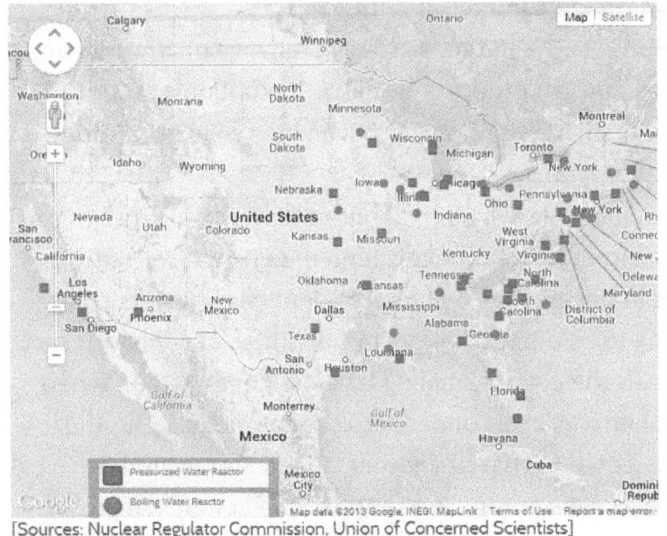

[Sources: Nuclear Regulator Commission, Union of Concerned Scientists]

1. Reduce the cost of construction and pay for the accident insurance without government subsidy.

2. Find and construct a permanent depository for high level nuclear waste

3. Reduce the cooling water requirements for nuclear power plants.

4. Convince the public that nuclear energy is sufficiently safe to live next to a nuclear plant in its neighborhood

5. Increase the nuclear EROI to be superior to renewable alternatives with reasonable life times.

In response to the favorable expectations during and after the Eisenhower administration, utilities in the United States built more than a hundred nuclear power plants located in the places shown on the map. These nuclear power plants provide today about 20% of our total electric power consumption use, which equals 8% of our total primary energy use. Most of these plants are now more than 30-years-old and, more than half of the currently operating plants will have to be replaced in the near future. The cost of a new nuclear power plant under current safety rules is about $8,000,000 per megawatt. Thus, replacing merely the existing plants in the U.S. which account for about 100,000 Megawatts would cost approximately 800 billion dollars, almost 5% of the 2013 U.S. GDP.

The 1992 National Energy Policy Act was the last piece of legislation signed by

President George H.W. Bush. It required that the nation's utilities be restructured. To comply with this legislation, the U.S. electric industry underwent major changes in its structure and operation. NCSL asked me to determine what opportunities solar technologies could offer for a restructured utility system. I had the opportunity to discuss this matter with many state legislators and review the past efforts of introducing solar energy into the energy system of various states. Experience in various states showed that solar and other renewable technologies can find a place in restructured utility systems. Also, several policy tools had been successfully integrated by a number of states in their restructuring plans. In order to enhance opportunities for solar technologies, the state governments successfully used primarily the following five legislative tools:

- Renewable Portfolio Standards (RPS)
- Net metering
- Rebates and Buy-Downs
- Tax Incentives
- Green Pricing

RPS provides states with a method to increase renewable energy generation in a cost-effective and administratively efficient manner. An RPS requires electric utilities to supply a specified amount of consumer load from eligible renewable energy sources. The utility can decide what technology is best suited and most cost effective for the locality.

I have personally taken advantage of three of these enhancement tools when I installed a PV system on my house in 2008. The previous year the voters in Colorado passed a referendum that introduced in its state constitution a mandatory renewable portfolio standard. Under this standard, the major utilities in the state are required to increase the percentage renewable in the total energy supplied by an ever increasing percentage up to a total of 20 % by the year 2020. Furthermore, utilities are required to purchase excess electricity produced by a private photovoltaic or wind system and to measure that amount by installing a watt meter that can indicate the amount used by the customer, as well as that supplied by the customer to the electric grid. In the end, a customer would only be billed for the net amount of energy used during the year. This is called NET METERING.

In order to achieve the required percentage of the RPS, the utility company for my neighborhood, the local utility Xcel Energy, provided rebates to any of its customers that installed a PV system on their home. I installed a 3 kW PV system on the roof of my garage in 2008. The total cost of that system was about $26,300 and $13,500 of the total was rebated by the utility Xcel and $500 was rebated from the Colorado sales tax. In addition, the federal government provided an exemption of up to $2,000 from the income tax during the year that the PV system was installed.

Unfortunately, the tax charged by the Boulder City and County was not waived, and amounted to approximately $1,500 out of the net cost of $12,273 for the PV system. Assuming that my system would live up to the expected electric power generation, it would repay the investment at the current $0.15 kW-hr cost for electricity in about 12 to 14 years. Thus, my own experience demonstrates that a distributed solar PV system in favorable locations can be a reasonable investment for an average person and certainly contribute to an overall scheme of introducing solar energy in a reasonable amount of time. At the time I installed my system PV was still too expensive to be competitive without substantial rebates, but since then the cost has dropped substantially.

President Clinton's cornerstone in his efforts to reduce our reliance on imported oil during his administration (1992 – 2000) was the Partnership for a New Generation of Vehicles, PNGV (http://www.pngv.org/). The PNGV was a cooperative research program between the U.S. government and major auto corporations, eight federal agencies, national laboratories, universities, and the United States Council for Automotive research. The PNGV challenge was "build a car with up to 80 mpg of the level of performance, utility, and cost of ownership that today's consumers demand."

By the year 2001, the Partnership, which had been formed in 1993, was on track to achieve its objectives. I was present at one of the last project reviews held in Golden which provided an insight into the progress of the PNGV program. According to the National Research Council review of the PNGV program in 2001, as well as the U.S. DOE publications of the same year, the PNGV program "overcame many challenges and has forged a useful and productive partnership of industry and government participants." It resulted in "three concept cars that demonstrate the feasibility of a variety of new automotive technologies." Both Ford and Chrysler had created working concept vehicles that achieved at least 72 mpg, while GM created the 80 mpg Precept. Unfortunately, although the PNGV was close to providing the country with a fuel efficient vehicle in accordance with its objectives, the program was cancelled immediately after the election of George W. Bush by his administration in 2001 at the request of automakers.

The Bush Administration discontinued the PNGV program as soon as it came to power and started instead a research program entitled "Freedom Car" which was designed to utilize fuel cells powered by hydrogen as the automotive fuel of the future. But hydrogen is not a fuel; it is merely an energy carrier that has to be created from some other basic energy source. Thus, soon after the announcement of the new initiative, bolstered by more than a billion dollars of research support for new proposals, Professor West and I demonstrated in several publications that the promise of the hydrogen economy was based on erroneous premises, as well as a lack of understanding of the thermodynamic disadvantages related to the inefficiencies in producing hydrogen and converting it into an energy carrier. In addition, the implementation of using hydrogen would require an entirely new infrastructure for hydrogen storage and transportation, estimated to cost of the order of half a trillion dollars. The concept of a hydrogen economy has now been virtually designated as no more than a vague hope for the future. As a wag put it, "hydrogen is the fuel of the future and it will always remain so."[3]

During the final years of NCSL, I devoted my attention almost exclusively to the transportation sector. Although everyone was aware of the problems caused by pollution of automobiles and the need to import petroleum which resulted in an enormous transfer of wealth to petroleum producing countries as well as an unfavorable trade balance, as well as the dangers to national security, the federal government was not willing to tackle this problem effectively. The only state that attempted to deal with the problem while I was at NCSL was California. In an effort to clean up air pollution, the California Air Resources Board (CARB), in 1990, adopted the ZEV plan which required an ever-increasing percentage of all new vehicles sold in the

[3] Kreith, F. and West, R.E. Fallacies of a Hydrogen Economy: A Critical Analysis of Hydrogen Production and Utilization. *Journal of Energy Source Technology*. Vol. 126. Dec. 2004.

state to be zero emission. The percentage was to increase to 10% by 2003. Since the only zero emission vehicle at the time was the electric vehicle, automobile manufacturers were forced to produce and try to sell, under this mandate, electric vehicles. NCSL provided information to CARB indicating that the technology was not yet ready because the lead acid batteries were too heavy, could not be charged fast enough, did not provide adequate mileage and there was no infrastructure to support the EVs. The reaction of a majority of Californians was "Hydrogen cars are great – I hope my neighbor buys one ".Therefore the mandate was not realistic. Our efforts to dissuade CARB were not heeded and a few years later the zero emission mandates was revoked because the public was unwilling to purchase electric vehicles. This was a good example that was noted by many other states, demonstrating that unless a technology is ready for the marketplace, a mandate will fail and should not be used in the hope of promoting the development of that technology. Moreover, not only the technology, but also the infrastructure must be in ready for a new transportation system to take hold. These lessons were important for other states and NCSL was able to widely disseminate them. However, NCSL subsequently went beyond a mandate for all- electric vehicles and, a 1996 publication entitled "Hybrid Electric Vehicles: Options for State Policymakers," proposed promoting electric hybrid plug-in vehicles by adopting appropriate incentive programs. This recommendation was premature, but today it has been adopted as federal policy and has been successful in promoting electric vehicle technology. In another publication of 1996, I prepared an overview of alternative fuels for the Science, Energy, and Environmental Resource Committee at the NCSL Assembly of State Issues. This publication stressed the future needs for alternative fuels and what states could do to promote their environmentally sound development. So, all in all, I believe that the efforts made by NCSL towards sustainable energy transportation policies were on the right track. Although they did not immediately bear fruit, they did lay the groundwork for future developments that today have resulted in legislation that has made important steps in promoting a more secure energy future.

The success of the NCSL efforts to help introduce legislation to facilitate the growth of sustainable energy is demonstrated in the two figures below. The first shows that today 36 states have introduced legislation for portfolio standards with requirements ranging requirements to up to 30% of energy production from renewable sources within the next five or ten years. The other successful policy measure is net metering. It is a part of the energy requirements in as many as 43 states, as shown in the figure below.

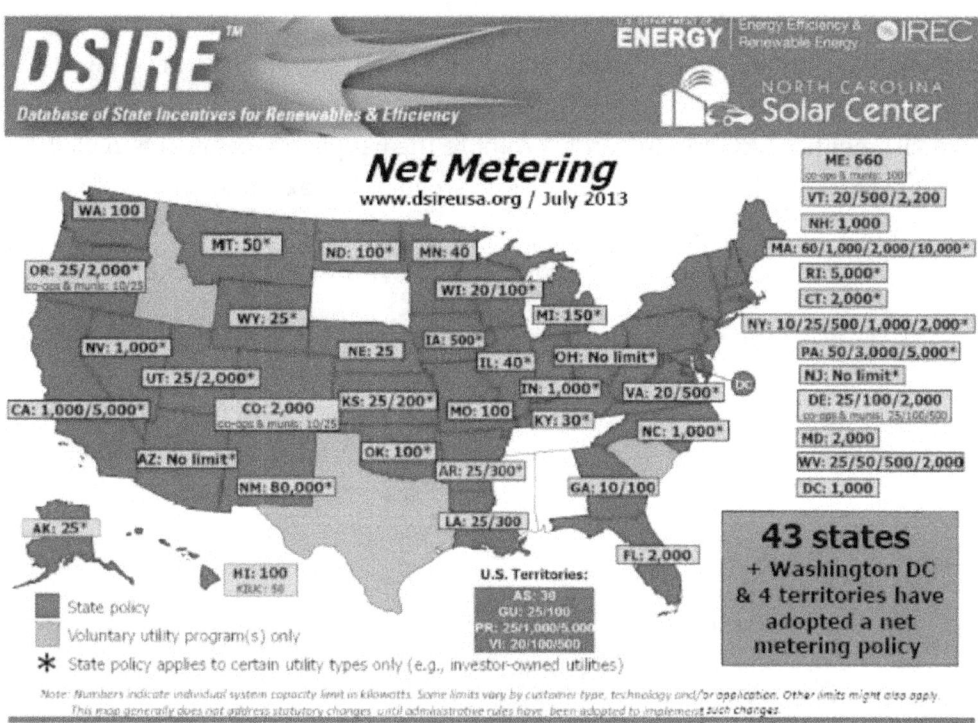

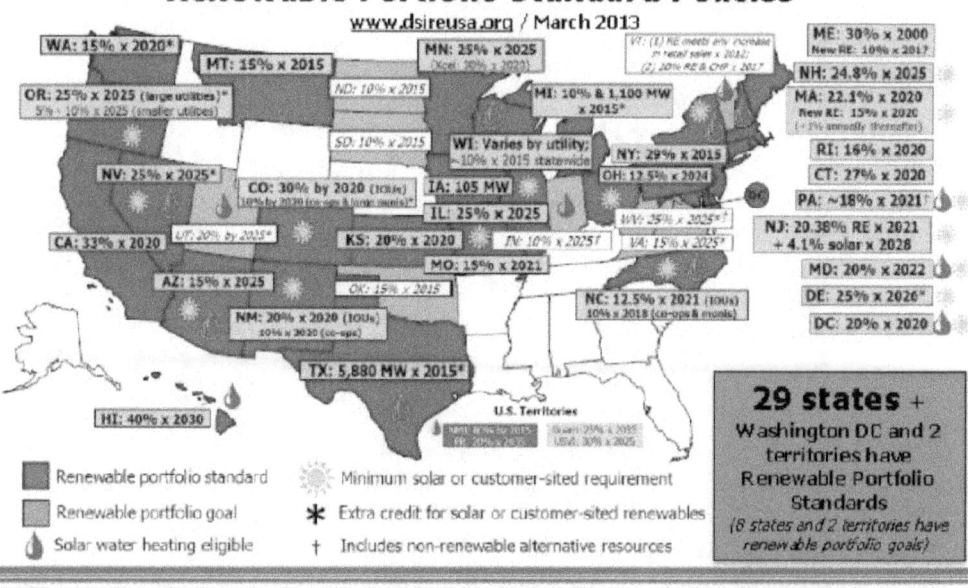

During my tenure at NCSL, I received recognition from various professional societies for my writing, teaching, and public service. But the most significant expression of appreciation for my work came unexpectedly in 1999 when the American engineering community expressed its appreciation of my efforts in promoting renewable energy to people in all the state governments in the United States by awarding me the Washington Award. This was an enormous and unexpected honor. Previous recipients include famous engineers such as the aviation pioneer

Orville Wright, the automobile giant Henry Ford, the engineering production pioneer Lillian Gilbreth, Admiral H.G.Rickover, and the astronaut Neil Armstrong. The Washington Award is "an honor conferred upon an engineer by fellow engineers for accomplishments which preeminently promote the happiness, comfort and well – being of humanity." The citation on my award reads: "For exceptional accomplishments in renewable energy conversion and conservation and for contributions to public policy which promote a better future for all people."

IN HONOR OF

Dr. Frank Kreith, Professor Emeritus of Engineering
University of Colorado, Boulder

TUESDAY EVENING
FEBRUARY 18, 1997
UNION LEAGUE CLUB OF CHICAGO
65 W. JACKSON BOULEVARD

AN HONOR CONFERRED UPON AN ENGINEER BY FELLOW ENGINEERS FOR ACCOMPLISHMENTS WHICH PREEMINENTLY PROMOTE THE HAPPINESS, COMFORT AND WELL-BEING OF HUMANITY.

THE COMMISSION
American Society of Civil Engineers; American Institute of Mining, Metallurgical and Petroleum Engineers; American Society of Mechanical Engineers; American Nuclear Society; Institute of Electrical and Electronics Engineers; National Soceity of Professional Engineers and Western Society of Engineers.

PREVIOUS WASHINGTON AWARD RECIPIENTS

1919 Herbert Hoover	1942 William Lamont Abbott	Herbert Payne Sedwick 1960	Dixy Lee Ray 1978
1922 Robert Woolston Hunt	1943 Andrey Abraham Potter	William V. Kahler 1961	Marvin Camras 1979
1924 Arthur Newell Talbot	1944 Henry Ford	Alexander Crawford Monteith 1962	Neil A. Armstrong 1980
1925 Jonas Waldo Smith	1945 Arthur Holly Compton	Philip Sporn 1963	John E. Swearingen 1981
1926 John Watson Alvord	1946 Vannevar Bush	John Slezak 1964	Manson Benedict 1982
1927 Orville Wright	1947 Karl Taylor Compton	Glenn Theodore Seaborg 1965	John Bardeen 1983
1928 Michael Idvorsky Pupin	1948 Ralph Edward Flanders	Augustus Braun Kinzel 1966	Robert W. Galvin 1984
1929 Bion Joseph Arnold	1949 John Lucian Savage	Frederick Lawson Hovde 1967	Stephen D. Bechtel, Jr. 1985
1930 Mortimer Elwyn Cooley	1950 Wilfred Sykes	James Brown Fisk 1968	Mark Shepherd, Jr. 1986
1931 Ralph Modjeski	1951 Edwin Howard Armstrong	Nathan M. Newmark 1969	Grace M. Hopper 1987
1932 William David Coolidge	1952 Henry Townley Heald	H. G. Rickover 1970	F. James McDonald 1988
1935 Ambrose Swasey	1953 Gustav Egloff	William Littell Everitt 1971	Sherwood L. Fawcett 1989
1936 Charles Franklin Kettering	1954 Lillian Moller Gilbreth	Thomas Otten Paine 1972	John H. Sununu 1990
1937 Frederick Gardiner Cottrell	1955 Charles Erwin Wilson	John A. Volpe 1973	Frank Borman 1991
1938 Frank Baldwin Jewett	1956 Robert E. Wilson	John D. deButts 1974	Leon Max Lederman 1992
1939 Daniel Webster Mead	1957 Walker Lee Cisler	David Packard 1975	William States Lee 1993
1940 Daniel Cowan Jackling	1958 Ben Moreell	Ralph B. Peck 1976	Kenneth H. Olsen 1994
1941 Ralph Budd	1959 James R. Killian, Jr.	Michael Tenenbaum 1977	George W. Housner 1995
			Wilson Greatbatch 1996

THE WASHINGTON AWARD COMMISSION
CORDIALLY INVITES THE MEMBERS AND FRIENDS
OF THE PARTICIPATING SOCIETIES TO THE

1997
WASHINGTON AWARD
BANQUET

NEIL A. ARMSTRONG
P.O. BOX 436
LEBANON, OH 45036

February 13, 1997

Dr. Frank Kreith
The Western Society of Engineers
53 W. Jackson Boulevard
Chicago, IL 60604

Dear Dr. Kreith:

I was delighted to learn of your election as the 1997 recipient of
The Washington Award of the Western Society of Engineers.

I am unable to be present to congratulate you personally, but I
hope that you will accept this note as a marginally acceptable
substitute.

Best wishes for an enjoyable evening.

Sincerely,

Neil A. Armstrong

NAA:vw

congratulating me. Among all of these letters, there was one that I particularly cherished. It came from the astronaut, Neil Armstrong, who was the first man to walk on the moon in 1969. His footprints and saying, "That's one small step for man, one giant leap for mankind," have become memorable. He had received the Washington Award in 1980.

What I remember most vividly many years after leaving NCSL is not the recognition I received for professional contributions, but rather the social interactions with state legislators and their staff at the annual NCSL conferences in various parts of the country. It was fascinating to see and talk with the country's decision makers who are serving in state legislatures and listen to their different approaches to promote a secure energy future. In my interactions with individual legislators, in conferences as well as in oral testimonies, I found all of them to be anxious to learn more about energy. At the same time, I observed a deep gulf between their understanding of the basic thermodynamics and engineering related to sustainable energy and the enormity of the issues facing them. It is a challenge for educational institutions to improve communication between technically trained people and intelligent individuals from other fields who lack that background. There are pitifully few engineers serving in legislatures and in congress. I believe that the challenge to bridge this gulf in knowledge can be met by interdisciplinary courses that bring engineering majors in energy and students from other colleges in contact. I embarked on this challenge in the fall semester of 2008 by offering an interdisciplinary honors course on sustainable energy for the first time at the University of Colorado.

CHAPTER 36—RETURN TO ACADEMIA, 2003–PRESENT

After writing my letter of resignation to NCSL I assumed that I would spend the rest of my life in true retirement. My search for a retirement hobby was cut short by a most unexpected, but welcome event. One of the professors of mechanical engineering at CU, Dr. Roop Mahajan, called me one day and asked if I would like to teach or co-teach a course in the department. He knew of my background in heat transfer and solar energy and told me it would be a shame if I would not share my experience with students. I was really taken aback because after my forced retirement in 1979 I had not set foot in the engineering college for almost 20 years. But this was an offer I could not refuse and I offered to jointly explore opportunities for returning to some limited teaching. I would have really liked to teach heat transfer again, but realized that with class sizes on the order of a hundred and fifty students it was not wise to accept such a responsibility. However, the department also needed experienced engineers to help in its mandatory Senior Design Course which required supervising a one-year-long

 industry-sponsored design project with teams of 4 to 6 students. This suited me better than a course with 100 plus students and I accepted responsibility for two teams. By officially returning to teaching on the campus in 2001, I also had the great opportunity to interact with graduate students in renewable energy and thermal sciences and to continue research with my old friend, Ron West.

Another truly happy occasion was the celebration of my 80th birthday. My birthday occurred close to one of the annual meetings of the American Society of Mechanical Engineers and my friends and colleagues arranged for a birthday celebration. This was

also an opportunity to meet again my old friend and long term professional colleague Professor D. Yogi Goswami who is standing next to me in the photograph. Yogi is a most unusual person in many ways. He is a devout Hindu who has lived in accordance with the principles of his religion all his life. He is a strict vegetarian, always aware of his environment, and seems to have a smile for everyone, no matter what the occasion. He is a firm believer in sustainable energy which seems to fit his religious beliefs. He also has both the academic credentials and the financial ability to obtain funding for his work.

Yogi and I met at a meeting in Hawaii in March 1987 and formed a close professional friendship. We did not often personally interact with each other, but over the years have cooperated on some important projects such as the monumental Handbook of Solar and Renewable Energy and the third edition of The Principles of Solar Engineering with Jan Kreider. We are now engaged in the second edition of our handbook and are hoping that it will become a useful tool for all of those engineers and decision makers who are planning for a sustainable energy future.

Supervising a team of students in their year-long design projects in their senior year gave me a unique opportunity to interact with students on a more intimate basis than a lecture course, and also brought me in contact with some real-world energy problems. The projects for this course were real world problems proposed by industry or national laboratories. Each of the projects was supervised by the industrial sponsor and an academic leader. Among the topics that I had the opportunity of supervising was the design of a national test bed for determining the thermal performance of energy saving windows. That performance criterion is commonly known as the R-value, and is an indicator of how effective a window's resistance to heat loss is. The project was sponsored by the National Renewable Energy Laboratory, my old stomping ground, and brought me in contact with many of my previous colleagues. The team did an outstanding job and our design was recommended to become the national test bed for home builders.

In another research project, a Japanese company by the name of Ricoh that had acquired the printer division of the American company IBM, asked that we investigate the potential of conserving energy by drying the printed output from one of their large printing presses by means of a laser with a frequency tuned to the colors of the print. This was an exciting project, but it turned out that, although the idea was technically feasible, no lasers of the required frequency range were available commercially. These lasers had been used for military applications and were potentially quite dangerous to the eyes. However, the team was able to provide some ideas for energy conservation which were eventually adopted by Ricoh. For another project, we were asked to investigate the possibility of cooling high energy electronic devices by means of fluids flowing through nano-channels at the back of the instruments. This is a continuing challenge nowadays, as electronic devices are becoming smaller in size, while the internal energy generation increases as more and more functions are incorporated.

But, the most interesting of the projects was proposed by the Boeing Aircraft Company. The goal of this project was to develop and build a prototype solar energy generation system that could be sold by an outlet such as Home Depot or Sears Roebuck. The solar collector was to be instrumented in such a way that it could be set on any roof and would orient itself to face the sun, in order to concentrate the solar rays onto a high-efficiency photovoltaic cell. During the preceding decade, the efficiency of PV cells had increased enormously, as shown below.

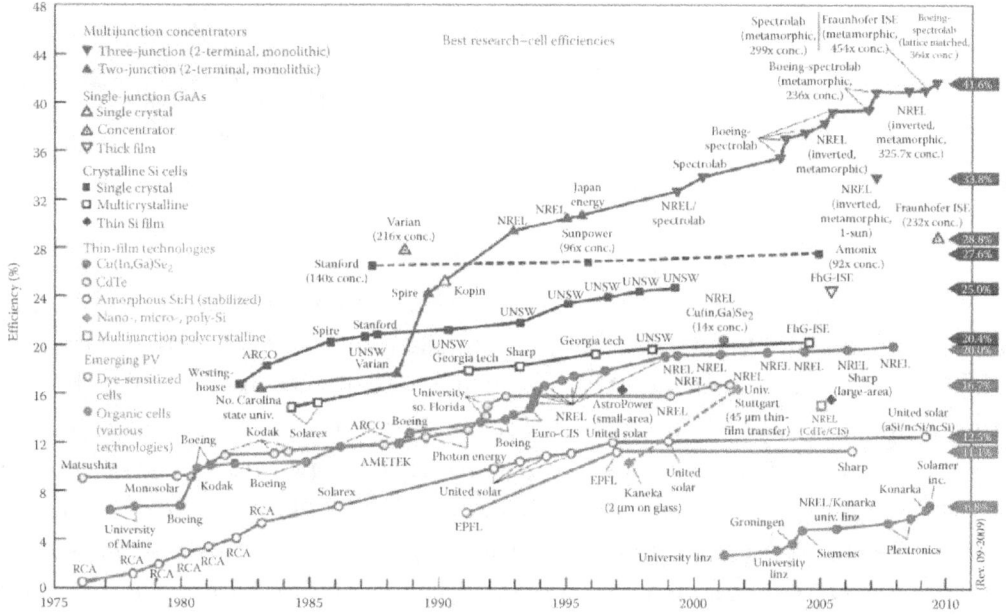

FIGURE 6.26
Best research-cell efficiencies. (From Jayarama Reddy, P., Cell efficiencies, in *Science and Technology of Photovoltaics*, 2nd edn., CRC Press, Boca Raton, FL; Kurtz, S., Opportunities and challenges for development of a mature concentrating photovoltaic power industry, NREL, NREL/TP-520-43208, revised November 2009. With permission.)

One of the most active companies in PV efficiency improvement was Spectrolab, a small startup located in Boulder. Boeing did not simply acquire the patent rights to this new PV cell, but simply bought the company in order to position itself for mass production of this new concept. Photovoltaic is not a field for mechanical engineers, and we were supplied with the cell, but asked to devise the optics to concentrate the solar rays and develop the software that would let the concentrator track the sun. The team made good progress during a year of R&D, and Boeing asked that the project be continued through a second year. Although the second year team had to first learn from the experience of the previous team, they again made excellent progress. It was very fortunate that in the second year team one of the members was a fifth-year student with a combined ME and electrical engineering degree. That student had also served as an intern at the National Center for Atmospheric Research (NCAR), where he had become knowledgeable of understanding celestial motion and was therefore able to design the tracking software for the Boeing device. At the end of the year, the team was ready to ship a prototype as shown below to Boeing.

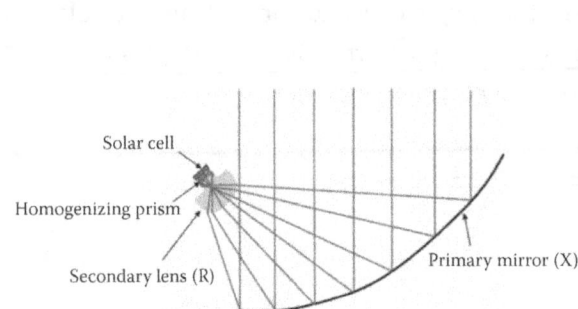

However, to our dismay, the Boeing project manager told us that Boeing decided to discontinue the project and had sold the part of the company that was instrumental in its development. When I asked why this decision was taken, the project manager told me that upper management had decided to discontinue the solar effort because they were afraid that, if this photovoltaic generator should malfunction due to its untried optics, it would reflect badly on the home company whose main business was selling airplanes. Management therefore decided not to take a chance on a novel renewable energy product, although initially they thought that this device would give the company a head start in a new and promising field for the future when the airplane business would wane as a result of continuing increase in the cost of jet fuel for the large Boeing planes. The fate of this project certainly gave me an insight into the problems that imaginative renewable energy projects face when evaluated by a profit-oriented organization in the real world.

CHAPTER 37— ASME FRANK KREITH ENERGY AWARD, 2005–2015

In 2005, I received a most pleasant surprise. The American Society of Mechanical Engineers International, for whom I had served for many years as its Legislative Fellow at NCSL, as well as Founding Editor of the ASME Journal of Solar Engineering and author of articles on sustainable energy, recognized my contributions by establishing an Energy Award in my name.

The description of the award states that the award is to be given to an individual who has made significant contributions to society that will lead to a sustainable energy future. The person honored with this award is also invited to give a

Frank Kreith Energy Award

The Frank Kreith Energy Award was established in 2005 to honor an individual for significant contributions to a secure energy future with particular emphasis on innovations in conservation and/or renewable energy. Contributions may be through research, education, practice or significant service to society that will lead to a sustainable energy future.

The Award was established by the Solar Energy and Advanced Energy Systems Divisions to honor Dr. Frank Kreith's contributions to renewable energy and heat transfer.

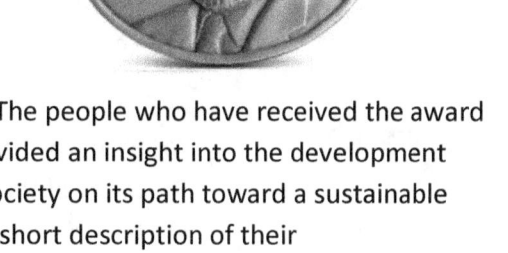

special lecture at the annual ASME meeting. The people who have received the award so far have given important lectures that provided an insight into the development renewable energy that will hopefully assist society on its path toward a sustainable energy future. Photos of the awardees and a short description of their accomplishments follow.

Roland Winston is a leading figure in the field of nonimaging optics and its applications to solar energy, and is sometimes termed the "father of non-imaging optics". He is the inventor of the compound parabolic collector (CPC), a breakthrough technology in solar energy. He is also a former Guggenheim Fellow, past head of the University of Chicago Department of Physics, a member of the founding faculty of University of California Merced, and as of 2013, head of the California Advanced Solar Technologies Institute. He holds more than 25 patents, chiefly related to solar energy, and has been figuratively said to have a "patent on the sun".

Yogi D. Goswami is the John & Naida Ramil Professor and the Co-director of the Clean Energy Research Center at the University of South Florida. He conducts fundamental and applied research on Solar Power and various other applications of solar energy. Yogi has served as the President of the International Solar Energy Society (ISES), a Governor of ASME-International, Senior Vice President of ASME and President of the International Association for Solar Energy Education. Some of his inventions have been commercialized and are available worldwide.

Ari Rabl was Senior Scientist at the Centre Energétique et Procédés of the Ecole des Mines in Paris until his retirement in 2007. He is continuing as consultant for government and industry, and in particular for the Ecole des Mines in collaboration with his old colleagues of the Centre Energétique et Procédés. After many years of research on energy and energy conservation he shifted the focus of his work in 1992 to the assessment of environmental impacts of energy technologies.

Robert Socolow is Professor Emeritus of Mechanical and Aerospace Engineering at Princeton University. His current research focuses on global carbon management and fossil-carbon sequestration. He is the co-principal investigator (with ecologist, Stephen Pacala) of Princeton University's Carbon Mitigation Initiative (CMI), a fifteen-year (2000-2015) research project supported by BP (and formerly by Ford). Under CMI, Princeton has launched new, coordinated research in environmental science, energy technology, geological engineering, and public policy.

Byard D. Wood is Professor and Head of Mechanical & Aerospace Engineering at the University of Utah. He has been the principal investigator for 56 sponsored research projects with total funding of $23 million from Federal, State and Local government agencies, Utilities, and Industry. Major projects include: Project Director for multidisciplinary research team in Biofuels & Bioproducts using phototrophic microbes such as algae; PI for a multi-organization team project to develop a hybrid solar lighting system using fiber optics to transport solar light to interior spaces.

Ann Marie Sastry is the President and Chief Executive Officer at Sakti 3. Sakti is commercializing a breakthrough, high performance, low cost and intrinsically safe solid-state battery technology. Prior to leading Sakti3, Sastry was the Arthur F. Thurnau Professor of Engineering at the University of Michigan. Tenured and promoted early, she founded and led two research centers in batteries and bioscience, and a global graduate program. She has received several of the highest technical honors in her field, including the 2007 ASME Gustus Larson Award and the NSF's PECASE.

Jane H. Davidson is Professor and Ronald L. and Janet A. Christenson Chair in Renewable Energy, and Director of Graduate Studies at the University of Minnesota. Jane Davidson works on solar systems to heat water, to warm and cool buildings, and to provide heat for industry and agriculture focusing on improving the thermal behavior of these systems and development of new materials to reduce cost and to provide long-term storage for cold climates. She has served as Editor of the ASME Journal of Solar Energy Engineering and is the recipient of the American Solar Energy Society Charles Greeley Abbot Award.

James E. Smith is a professor in the Department of Mechanical and Aerospace Engineering at the Benjamin M. Statler College of Engineering and Mineral Resources at West Virginia University (WVU) and the founder of WV PRIME. Throughout his more than 40 year career, Smith has been the principle or co-principle investigator for research projects sponsored by federal agencies, such as the Department of Energy and the Department of Defense.

CHAPTER 38—TEACHING SUSTAINABLE ENERGY, 2005–2010

"If you surrender to the wind, you can ride it." – Toni Morrison

In December of 2005, I celebrated my eighty-third birthday without any more future professional plans. However, Dr. Roop Mahajan, who had been promoted to Acting Dean, visited me in my office and said "Frank, I know you may not want to teach an entire course, but maybe you would be interested in developing a course in sustainable energy. We would like to develop such a course in the department and can give you an assistant who can carry much of the load". That idea excited me because I had been aware that despite the recognition of the need for a transition to a non-fossil energy system, the educational establishment at CU did not have any suitable courses to prepare engineers for that future assignment. The previous year, ASME had elected me to Honorary Member, which is the highest accolade that the society can bestow on one of its members and I was no longer interested in sponsored research and more publications. But the idea of developing a course in sustainable energy seemed very appealing. I began by looking for a suitable textbook but could not find one that would meet the objectives set for the course I wanted to teach. I therefore began to collect material that would be suitable for a course at the senior or first year graduate level for engineers that had a background in thermodynamics and computer technology. Looking around the staff in the department, I did not find anyone on the teaching staff suitable to assist me in the development of this course. However, one of the untenured research assistants, Dr. Mike Hannigan, had shown an interest in pollution control and teaching thermodynamics. I therefore approached him and he agreed to assist me in the development of the first sustainable energy course at the university. We were given a small budget to hire teaching assistants for some library research and announced the presentation of the first sustainable energy class for the fall of 2006. The description of the course began with this overview "Society's dependence on the fossil fuel sources which began during the last half of the 19th century is no longer sustainable. It is therefore necessary to plan for a transition to a broader energy mix that includes a variety of renewable sources, including wind, biomass and solar. The goal of this course is to provide our engineers with the background necessary to participate in this transition." The course was well attended and launched me on a new professional goal in the waning years of my professional life: the development of a text for a course in sustainable energy systems.

In retrospect, I realized that the description of the course had omitted an important topic, namely energy conservation. Although the matter of conservation was not mentioned in the original course description, it was included as an important facet of the course and added to the course description for the following years. The initial sustainability course was offered by the ME Department as a technical elective for any engineering students with prerequisites in thermodynamics. It was not intended to be interdisciplinary. After 3 years, when the enrollment had grown to more than 40 students, Dr. Hannigan took over teaching the ME course, and I started to look for a more suitable venue.

Many years ago I had been involved in the Honors Program at CU. This is an interdisciplinary program which limits class sizes to 15 students and accepts only students with a grade point average in the upper fifth percentile. I proposed to the director of the Honors Program my interest in developing an interdisciplinary course on sustainable energy, and he accepted the proposal. In 2007, I began to offer this course as part of the CU Honors Program curriculum. This course gave me an opportunity to teach a small class of outstanding students and utilize all of my past experience in energy and my commitment to aiding the transition to a renewable energy system that I believe to be unavoidable. I may not have the opportunity to observe and participate in the long-awaited rise of renewable energy on the national as well as international scene, but I hope that some of the students I have taught will prove to be helpful in the transition. One of them has already served as an energy advisor to the Governor of Colorado.

There were several outstanding guest lecturers who contributed to the course, including the late Professor Al Bartlett, who during his life has given hundreds of lectures all over the country about the danger of overpopulation to a healthy society in the future. Additional lecturers were Dr. Rita Klees, a consultant on water issues to the United Nations, Mr. Paul Norton, a former graduate student of mine who had worked with me at SERI and retired from NREL as a senior engineer, Dr. Isaac Garaway on leave from the Technion in Israel, Professor Susan Krumdiek from the University of Canterbury, New Zealand,Professor Jana Milford of CU Mechanical Engineering, Professor Mosef Krarti of CU Civil Engineering and of course my old friend Ron West. The interaction of all these people with the Honors students in class was an invaluable input to my perspective on sustainability. Some of the students continued to work with me on the book that was an outgrowth of the course. In lieu of a final examination student teams were asked to develop a scheme for sustainability in a country of their choice. Their work showed that energy sustainability is within reach of many countries. Also the local newspaper took notice of the development of the course and published an interview with me about the course.

University of Colorado professor Frank Kreith stands in his home in 2007.

Cliff Grassmick | Camera file photo

Students work on energy problems

By Laura Snider
Camera Staff Writer

Terese Decker was frustrated. After crunching the numbers again, reworking the problems, doing more research and running complex cost-benefit analyses, the senior at the University of Colorado created a plan that allows Spain to get an incredible 98 percent of its energy from renewable sources by the year 2030.

But she just couldn't squeeze out that extra 2 percent — representing the burning of fossil fuels — in any kind of reasonable, cost-effective way.

"That's a little disappointing," she told her classmates Tuesday.

Decker and her fellow "en-

See PROF, 5B

ra.com

Prof pushes students to focus on world issues

Continued from 1B

ergy minister," Josh Maynard, presented their semester's work on Spain to professor Frank Kreith this week. The students in Kreith's sustainable energy class have been working in pairs since August to write energy plans for a handful of countries, including Israel, Turkey, Thailand, France and Brazil.

The students, who are all a part of the honors program, took a comprehensive look at their individual countries, considering energy issues from every angle, including the country's natural resources, population density, arable land, conservation goals, quality of life, government structure, culture and wealth. The resulting plans took a stab at

being realistic paths toward sustainability.

"This is by far the best class I've taken," said Kyle Converse, a senior in mechanical engineering who created a plan for Brazil.

Converse credits Kreith, who first began teaching at CU a half-century ago (and has no apparent plans for quitting anytime soon), for the class's power. Kreith shared cake and coffee with his students this week to celebrate his 86th birthday.

Kreith recently returned to CU after leaving in 1978 to work for the Solar Energy Research Institute and then the American Society of Mechanical Engineers. During his career, he has advised state legislators, the United Nations, the U.S. Department of Energy and the U.S. Department of

State, and he worked on President Jimmy Carter's domestic energy review.

When Kreith walked into class Tuesday, he put down his wooden cane and got right to work checking in with his students. He knows them all by name and badgers the ones who still have missing assignments.

One student handed in some late work, but Kreith looked disappointed.

"I can't read it," he said. "I'll have to get a magnifying glass."

"I was trying to save paper," the student explained, "and energy."

Despite the play at Kreith's soft spot — conserving energy — it didn't work.

"Is that it?" Kreith said. "My eyes are more important."

The spring 2007 commencement represented both a beginning and an end for me. The May 11th ceremony officially announced the retirement of my first doctoral student, Dr. Jan F. Kreider, and the award of a PhD to the last of my doctoral students, Dr. Ming Ye, who was co-advised by me and Roop Mahajan for his PhD dissertation in mechanical engineering. Jan, after receiving his PhD, had become a Professor in the Department of Civil, Environmental, and Architectural Engineering and also the Founding Director of the University of Colorado's Joint Center for Energy Management. I have great respect for his professional achievements and was sorry to see him retire so early.

UCB prof's last Ph.D. student tosses cap, 1st one hangs it up

6/21/07

By Jefferson Dodge

Frank Kreith

For CU-Boulder Professor Emeritus Frank Kreith of chemical and biological engineering, the spring 2007 commencement represented both a beginning and an end.

At the May 11 ceremony, officials announced the retirement of Kreith's first doctoral student — and awarded a Ph.D. to his final doctoral student.

Kreith's first doctoral student, Jan Kreider, earned his Ph.D. in chemical engineering in 1973 and spent his faculty career at CU-Boulder, in the department of civil, environmental and architectural engineering. Ming Yi, the last doctoral student Kreith plans to mentor, received his Ph.D. in mechanical engineering at the commencement last month, after being co-advised by Kreith and Roop Mahajan of mechanical engineering.

Kreith, who was born in Vienna, was rescued from Nazi-occupied Austria via the kindertransport at the age of 15. He lived in England for a couple of years before moving to the United States, where he earned an undergraduate degree at the University of California at Berkeley and an M.S. degree from the University of California at Los Angeles. He received his doctorate from the University of Paris in 1965 while taking a year off from UCB.

Kreith, 84, said he came to Boulder in 1959 because he loved to ski, having learned the sport while traveling to school on wooden skis in Austria. "I skied for 70 years," he said proudly, adding that he just gave it up two years ago.

Kreith taught at UCB for 20 years, then spent a decade as chief of thermal research at the Golden-based Solar Energy Research Institute, now known as the National Renewable Energy Laboratory. Kreith also served as an American Society of Mechanical Engineers (ASME) legislative fellow at the National Conference of State Legislatures in Denver for 13 years, providing information on legislative issues related to energy and the environment. He has been ASME government coordinator for Colorado the past two years, helping develop legislation on renewable energy.

Four years ago, he decided to return to UCB to teach part time in mechanical engineering. Last fall, he taught a new course in the department on sustainable energy. "I came back to teach because I think it's so important that students become aware of how important energy will be in their lives," he said. "The two main themes in this presidential election are the Iraq war and energy independence. The two are related. We have to have some independence and get away from oil. If we don't, we are in trouble."

Kreith said he has been encouraged by recent developments in Colorado on the renewable energy front, such as efforts by the voters and Gov. Bill Ritter to increase the minimum percentage of power that utilities must produce from renewable energy sources.

He said the United States does not yet have a complete energy policy and Colorado can play a leading role in developing one, since the state has a broad range of energy storage and production options. "I'm hoping the University can participate in this, and that's why I'm back teaching," he said.

Kreith also remains active in research, having recently published a couple of articles and co-edited the *Handbook of Energy Efficiency and Renewable Energy*. Kreith and his first doctoral student, Kreider, have co-authored three books, and they just signed a contract to write a new book on the principles of renewable energy.

Kreith said that while he has mentored his final doctoral student, he plans to continue teaching "as long as the students don't throw me out. I love students, and I love the interaction. It keeps me young."

Silver & Gold Record

CU's faculty and staff newspaper 6/21/07

UCB | UCCS
UCDHSC

The return to full time teaching has been a truly satisfying experience. I realized how complex the transition to a sustainable energy system will be and working with young engineers to understand this challenge, has taken up most of my professional energy and writing efforts during the following years. The teaching efforts culminated in an invitation to give a plenary address for the 2011 ASME Conference entitled "Teaching Sustainability: A New Engineering Challenge." I had a full audience for my talk, but my challenge was not as widely accepted as I had hoped. Some of my colleges later told me that their administrations were reluctant to back a new topic for which they did not see much financial support in budgets from Washington.

President Barak Obama, who was elected in 2008, publicized the need for sustainability and announced his support for measures to promote renewable energy to contain global warming. From their words it seems that it has become obvious to political leaders that the transition to a sustainable energy system, although mostly an engineering function, is heavily dependent on their leadership and involves social issues, such as population, food, water, a clean environment, economics and finally politics. But the saying that "Actions speak louder than words" tells a different story.

In 2012, I had an interesting and also encouraging experience. I received an invitation from the U.S. Air Force Academy to give a talk on sustainable energy; this invitation was a surprise since I didn't expect the U.S. Air Force Academy to be very interested in the subject. But I accepted the invitation and on April 30, 2012 gave a talk to a group of cadets in the engineering program. The officer, who received me, explained to me that a course in sustainable energy could in the future become mandatory for engineering cadets interested in a career in the U.S. Air Force. I asked the officer, who also teaches at the academy, what prompted this topic to be considered as a requirement in the education of engineering cadets. His response was that "in our opinion energy security is a key to national security." I told him I could not agree more, but was surprised that a military school should be so far ahead of the U.S. Congress in recognizing the importance of energy. His snapped response was "well it's easier for us to be reasonable in technical matters since we don't have to run for office every two years."

I have also found it very gratifying that in the past few years the topics of sustainability and climate change have found increased attention, both in industry and academia. In the fall of 2013 the College of Engineering at CU renamed one of its buildings the Energy Center for Environmental Sustainability and the ASME Heat Transfer Division added Sustainability as an additional topic for its 75th Anniversary meeting in 2013 as shown below.

ASME 2013

Summer Heat Transfer Conference

7th Conference on Energy Sustainability

11th International Fuel Cell Science, Engineering & Technology Conference

Hilton Minneapolis
Minneapolis, Minnesota
July 14–19, 2013

In a similar vein the American Institute of Chemical Engineers named its 2013 Annual Meeting in San Francisco ENGINEERING a SUSTAINABLE FUTURE and introduced the meeting with the words "Sustainability, and the global challenges and opportunities it offers chemical engineers, will be a uniting theme of discussion …." The meeting also held a joint session with IChemE entitled "Energy, Water, Food, Materials: An interconnected Global Challenge." Since then the number of professional meetings and seminars devoted to sustainability has increased continuously.

An organization which has played a big part in my life and in the development of renewable solar energy is the American Solar Energy Society, commonly referred to as ASES. ASES is a grassroots organization started by a few solar energy enthusiasts. The society has been at the forefront of bringing the potential of solar energy to a

wide public audience and has also tried to influence political decisions helping to develop renewable solar energy. I have been a part of this effort of the organization since 1975. I was elected a member of the board of directors in 1979 and also served on the editorial advisory board of Solar Age, the highly successful publication of ASES for the general public where I provided guidance to the journal for many years until I resigned in 2011 to make room for new blood. I continued publishing articles on solar energy, and gave one of the plenary lectures at the Solar ASES conference in Denver. ASES expressed its appreciation for my contributions by awarding me the Charles Greeley Abbott Award in 1988 for "deep personal commitment to the development of solar energy" and in 2012, the Hoyt Clarke Hottel Award with the citation "For his innovative thinking that spurred many technology advancements. He has publically promoted solar energy for decades, while inspiring generations of young engineers". The award came for my 90th birthday and is one of the most cherished presents I have received in my life.

Our family celebrated my 91st birthday on December 28th as a combined occasion for my son Michael's 60th birthday and my daughter Judy's 53rd birthday. Many of my friends and colleagues came to the party held at a local Mexican restaurant called Casa Alvarez. Among the guests was one newcomer: Carrie, the seeing-eye dog for Bev Weiler. Along with my family, Bev has been and still is the most important person in my life. She graduated with my daughter Judy from the elite Colorado College with the intention of becoming a lighting expert for the theatre. Her ambition was unfortunately cut short when a hit-and-run driver ran into her on the bicycle and damaged one of her knees beyond repair. She changed her life's direction to become a computer expert and book editor for many publishers, including CRC Press and McGraw Hill, who published my books. As my ability to write by hand diminished, I dictated my books and articles. Bev then made drafts from my dictations that I could edit into a final product. The last thirty years of my professional life would not have been possible without her. But another disaster struck Bev later in her life when it was

discovered that she suffered from retinitis pigmentosa and would be slowly going blind. Despite her diminishing eyesight, Bev has continued to work by having an enlarged computer screen that made it possible for her to see the words. Without Bev's help in editing, this book would never have been published.

That year's birthday was the only time when all of my immediate family came to together in our home in Boulder and we could take a picture of the occasion.

One person was with us only in spirit at that birthday party: my late sister Susi. Susi had lead an interesting life at the forefront of the computer industry. She had married her sweetheart from their Berkeley student days, who turned out to be one of the foremost computer geniuses in the world. Susi raised four brilliant children with her husband, but virtually devoted all of her energy to supporting Glen's professional life. Susi died in November of 1997 and did not live to see her husband's national recognition. Glen Culler was awarded the National Medal of Technology, the nation's highest technology honor by President Clinton on March 14, 2000 at a ceremony in the East Room of the White House. Glen was cited for "pioneering innovations in multiple branches of computing, including early efforts in digital speech processing…".

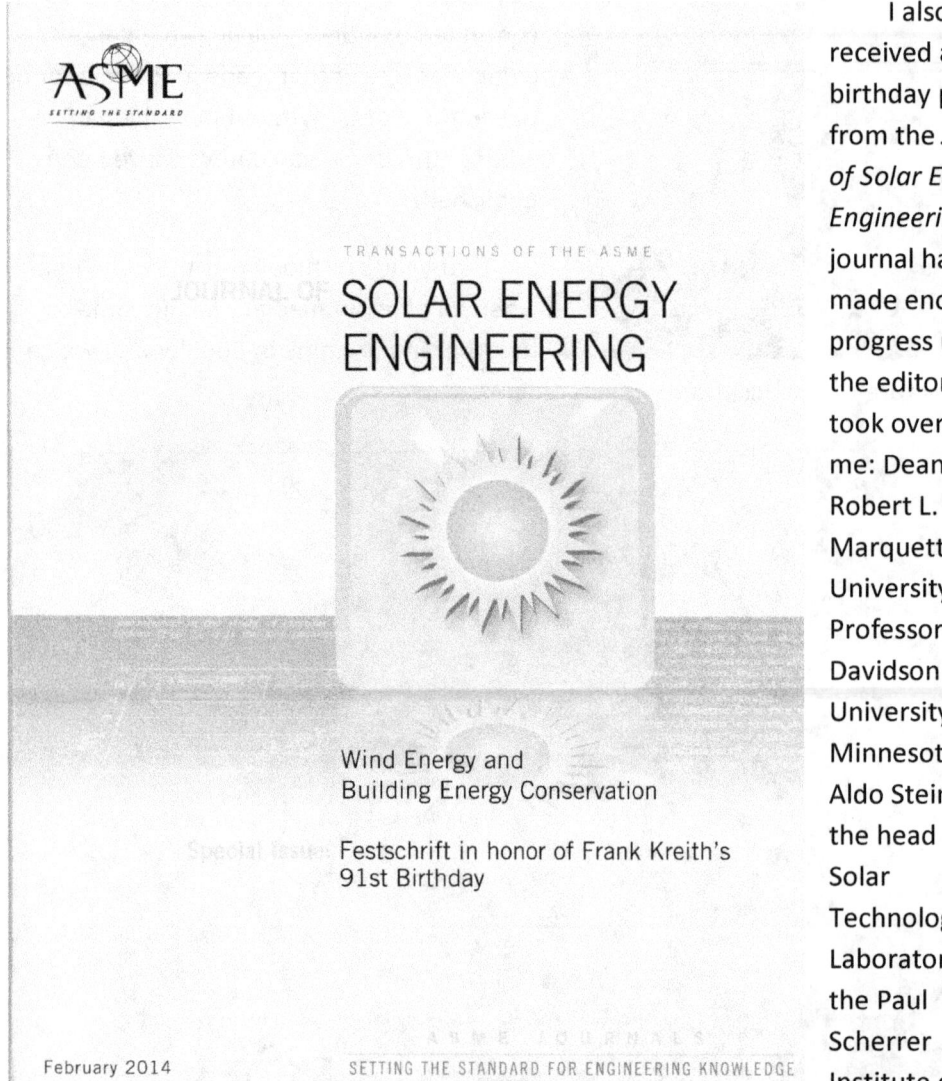

ASME
SETTING THE STANDARD

TRANSACTIONS OF THE ASME

SOLAR ENERGY ENGINEERING

Wind Energy and
Building Energy Conservation

Festschrift in honor of Frank Kreith's
91st Birthday

February 2014

SETTING THE STANDARD FOR ENGINEERING KNOWLEDGE

I also received a birthday present from the *Journal of Solar Energy Engineering*. The journal had made enormous progress under the editors that took over after me: Dean Robert L. Reid of Marquette University, Professor James Davidson of the University of Minnesota, Dr. Aldo Steinfeld, the head of the Solar Technology Laboratory of the Paul Scherrer Institute of Switzerland, and Dr. Gilles Flamant of the French National Centre for Scientific Research. Under their editorship, the journal has reached out to countries all over the world and has become the foremost peer reviewed journal in the field of solar engineering. The journal has officially added wind energy and building energy conservation to its coverage and met the ASME journal goal of setting the standard for engineering knowledge. The journal paid tribute to my contributions by issuing in February 2014, a special issue: Festschrift in honor of Frank Kreith's 91st Birthday. The contributors to this special issue included some of my best friends who had achieved eminence in the field of solar engineering.

CHAPTER 39—PERSPECTIVE ON FUNDING OF RENEWABLE ENERGY, 2013

"Don't tell me what you value, show me your budget, and I'll tell you what you value."
Joe Biden, Vice-President of the United States

It may not be appropriate to cite numbers when writing a autobiography. But I know of no other way of explaining why, given the verbal support for renewable energy by so many past presidents and the emotional preference of renewable sources for our future by the general public, so little progress has been made in actually transitioning from a fossil to a solar energy system. As Vice-President Biden says, one needs to look at monetary allocations in order to determine where the real values of our decision makers lie. I am therefore using some of the budgets for the past 40 years in order to demonstrate what the priorities in governments have been. From my experience, most legislators follow their own priority, "Do whatever it takes to get re-elected." This is the first law of politics and requires money from wealthy donors for the next campaign. And much of that money is in the hands of oil, coal and nuclear energy.

Until recently I did not follow the budget trends of DOE and their effects on the development of a sustainable and renewable energy system. However, my efforts to look at the current energy situation and how it could evolve into a sustainable energy future have prompted me to peruse the DOE budget over the past few years. These budget trends are shown in the figure below, based upon information from the American Association for the Advancement of Science (AAAS). It is apparent that soon after President George W. Bush was elected in 2000 the amount of money for energy conservation has consistently declined from about $710 million in the year 2000 to about $480 million in 2008. During the same time, yearly expenditures for coal R&D have increased from about $140 to $410 million and from $100 to $580 million for nuclear power. Expenditures for wind, geothermal, and biomass **combined** have remained at a threshold of about $200 million per year.

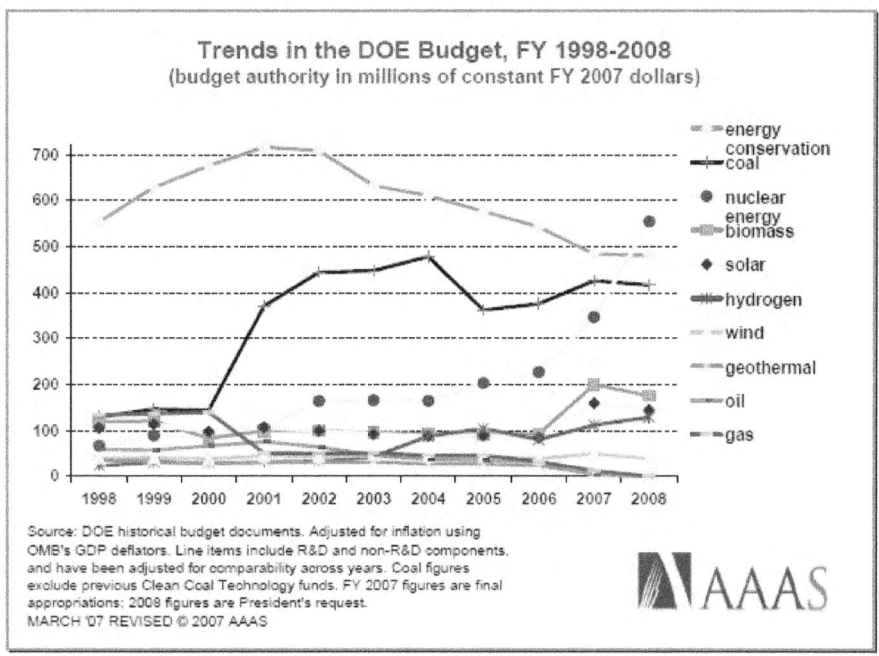

The situation regarding federal research and development has improved over the past six years, but still does not reflect the importance of sustainable energy for the country's future. However, between 2009 and 2014, the budget for DOE has increased from 11.5 to 12.5 trillion in constant FY2013 dollars. This is an increase of 13%, but still considerably less than the budget for defense which was 68 trillion dollars. In the same five years period, the R&D budget for DOD decreased by about 22%. But the Department of Energy budget is still only about 20% of the amount budgeted for the Department of Defense. [Reference AAAS report: Research and Development Series available at www.aaas.org/spp/rd]

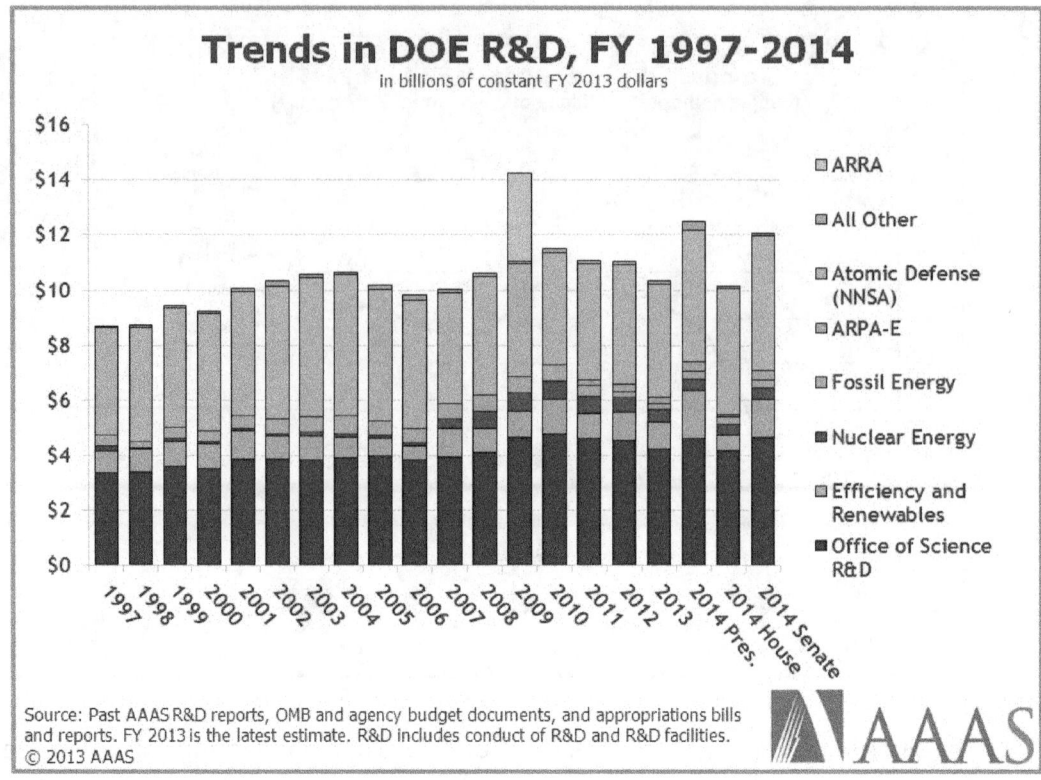

Trends in DOE R&D, FY 1997-2014
in billions of constant FY 2013 dollars

Legend: ARRA, All Other, Atomic Defense (NNSA), ARPA-E, Fossil Energy, Nuclear Energy, Efficiency and Renewables, Office of Science R&D

Source: Past AAAS R&D reports, OMB and agency budget documents, and appropriations bills and reports. FY 2013 is the latest estimate. R&D includes conduct of R&D and R&D facilities.
© 2013 AAAS

Table I-4. Major Functional Categories of R&D
(budget authority in millions of dollars)

	FY 2012 Actual	FY 2013* Estimate	FY 2014 Budget	Change FY 12-14* Amount	Percent	% of Total ('14)
Defense (050) 1/	78,717	72,132	**74,382**	-4,335	-5.5%	51.6%
Nondefense	63,807	61,080	**69,683**	5,876	9.2%	48.4%
Space (252)	10,762	10,116	**11,052**	290	2.7%	7.7%
Health (550)	32,617	31,144	**33,684**	1,067	3.3%	23.4%
Energy (270)	2,191	2,093	**3,211**	1,020	46.6%	2.2%
General Science (251)	10,168	9,717	**10,984**	815	8.0%	7.6%
Environment (300) 2/	2,301	2,276	**2,610**	309	13.4%	1.8%
Agriculture (350)	2,005	1,860	**2,182**	177	8.8%	1.5%
Transportation (400)	1,499	1,327	**1,512**	13	0.8%	1.0%
Commerce (370)	697	724	**1,964**	1,267	181.8%	1.4%
International (150)	269	256	**250**	-10	-3.7%	0.2%

The amount of funding allocated to energy per se in 2014 is even less encouraging than the funding for renewable energy within DOE . The FY2014 budget authority for all energy was about $3.2 billion, whereas the funding for space R&D was $11 billion and for health $33 billion. In comparison the amount budgeted for defense was $ 74.4 billion or about 52% of the total budget. In perspective, all of energy expenditure was only 2.2% of the R & D budget. Although the energy budget almost doubled in the period between 2012 and 2014, it still constitutes only a tiny fraction of the total amount spent on advancing all technologies as whole. Perusing the budgets, I am reminded of Nobel Laureate Sir George Porter's comment "I have no doubt that we will be successful in harnessing the sun's energy ...if sunbeams were weapons of war, we would have had solar energy long ago".

Given the absurd priorities that give only a small amount of money to support development of renewable energy, it is not surprising that so little progress has been made by government supported research in commercializing renewable energy technologies. The major advances that have been made are largely the result of enterprising ex-SERI and ex-NREL employees who struck out on their own or by start-up private investments, e.g. in wind energy,bio-mass derived fuels or passive solar heating.

But it is not only the overall picture that is disheartening. The National Renewable Energy Laboratory, the successor of SERI, was established as the flagship of Solar R&D. Yet, the level of its support suggests that it is more subject to political than scientific priorities. The 1980 reduction in force under President Reagan has already been discussed. But this action is not alone. For example, on December 21st, 2005 Kevin Humphrey reported that NREL is going to begin laying off as many as 100 of its scientists and researchers (11% of its staff) in response to drastic cuts in its 2006 budget. He added that NREL's budget would be $20 million less than its measly 2005 budget of $ 200 million. But this 11% cut pales when realizing that DOE's total renewable energy budget was cut by 35%. Even if some of these cuts were eventually reduced, these ups and downs in funding, dictated by political vagaries in DC, have a deleterious effect on the moral of the NREL staff and make it difficult to retain outstanding researchers.

Politics continued to raise its ugly head even more disturbingly when in early February the 2008 NREL budget was cut by $28 million. As a result, 32 researchers were laid off and outside contracts and sub-contracts were cancelled according to an article in nationmaster. com, 12/22/2008. But at about the same time, President George W. Bush gave a speech in which he deplored the nation's addiction to oil and announced that he would visit NREL on February 21, 2006. On February 20th, DOE restored $5,000,000 to NREL's budget and NREL officials stated that the lab would attempt to rehire the employees that had just been fired. The president subsequently said that the original budget cut was the result of "mixed signals." But there were no mixed signals when NREL announced in 2011 that between 100 and 150 positions were to be eliminated the following year. The blow was ameliorated by announcing a so-called "by-out "under which old timers willing to retire early would receive 2 months' salary for every year they had worked at NREL. This approach induced some of the most experienced researchers to retire and leave the laboratory. Their place was taken by younger less experienced workers at lower pay. In 2014 President Obama's $787 billion stimulus package included spending on renewable energy and energy efficiency programs, but according to DOE spokesman Chris Powers the way DOE must spend the money, it could restrict money going to NREL . (Stimulus leaves NREL in cold, *Denver*

Post, 1/29/2014).Anyone who has ever been in charge of planning research must realize that it is well-nigh impossible to run long-term programs effectively under such circumstances.

CHAPTER 40—*PRINCIPLES OF SUSTAINABLE ENERGY SYSTEMS*, 2013

"We simply must balance our demand for energy with our rapidly shrinking resources. By acting now we can control our future instead of letting the future control us" President Jimmy Carter

In the period between 2008 and 2013, while teaching the course on sustainable energy, I worked on a book entitled *Principles of Sustainable Energy Systems*. I tried to incorporate in this book most of the material I had previously published in some of my books and articles dealing with sustainable solar energy, but extended that material to include the wider recognition that the interaction of our energy needs also includes topics such as the population explosion, environmental degradation, resource depletion, and economics.

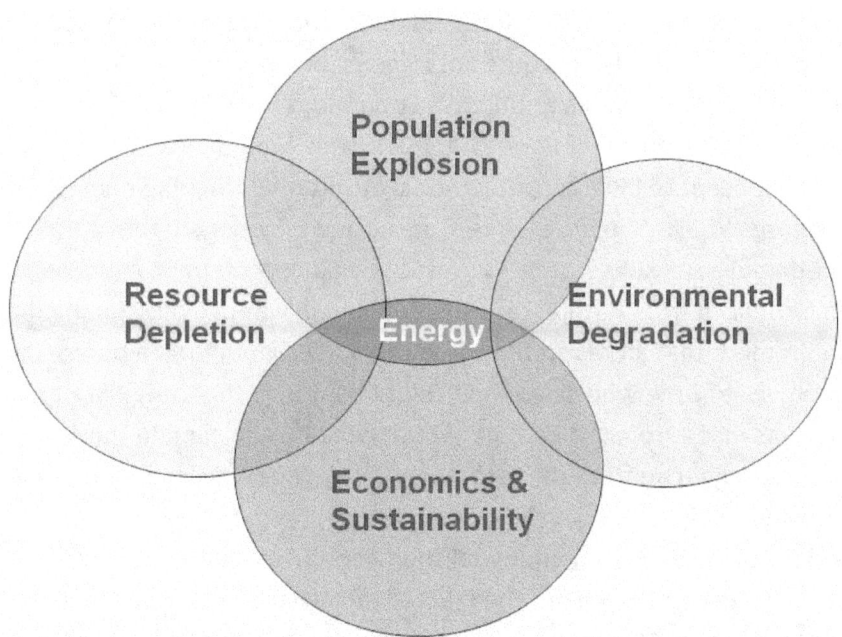

The book covers all of the sustainable technologies, but also has some novel topics necessary to plan for a transition from a fossil-based to a renewable energy system. In addition to the conventional economics, I included material on an emergent topic that is an imperative tool for planning a sustainable energy future: using energy instead of money in an analysis called net energy analysis (NEA). This approach gives the energy return on energy invested (EROI) instead of the monetary return on the money invested for any type of energy production or conservation system. I was introduced to this topic by the well-known economist Dr. Cutler Cleveland. Cutler and his colleague, Robert Costanza, published papers which have become important tools in understanding the difference and relation between fossil and renewable energy systems. The former needs a continuous supply of expensive fuel to operate, while the latter runs on a virtually unlimited and free supply. Supplying fossil fuel takes an ever increasing amount of energy and money investments.

Traditional thermodynamic courses rely on thermal efficiency defined as the ratio of energy output to energy input as the evaluation criterion for an energy system. But this approach does not take into consideration the amount of energy that is required to build, operate and decommission the system. For instance, if a system with 50% operating efficiency would take more energy to build than it can output during its lifetime, the thermodynamic efficiency alone will not be a good indicator of its value. But the EROI concept would ask what the ratio of energy output of the system is during its lifetime divided by the energy input to build and run the system. This ratio is called EROI and in equation form is shown below:

$$EROI = \frac{Energy\ Output\ over\ lifetime\ of\ system}{Energy\ input\ to\ build\ and\ run\ the\ system}$$

Viewing a solar system from this perspective shows that the concept of some solar enthusiasts who claim that solar energy is free and has no environmental impact is not correct. For example, if a solar collector puts out a thousand energy units during its lifetime but required 200 energy units to build and operate, the EROI would be 5 to 1. But if the output were 500 with an input of 1000 the system would be an energy hog, not a producer. The concept of EROI caught national attention during the energy crisis when President George W. Bush in order to replace oil and encouraged the development of ethanol as its substitute and provided an incentive of 54 cents per gallon of ethanol produced from natural resources. In Brazil, where a dictatorship had imposed the requirement that all gas stations also provide ethanol from a pump, the fuel was made from sugar cane which has an EROI of about 6. In the United States, however, ethanol was made from the kernels of corn because the climate is not conducive to growing sugar cane and the technology to use cellulosic material (that is the cob of the corn)was only under development, but not commercially available. When the Cornell biologist David Pimentel published an article claiming that when ethanol is produced from kernels of corn the EROI of the ethanol was less than one, the U.S. congress tool notice. According to this analysis, more energy would be needed to produce the equivalent energy output available from ethanol and therefore it would be a waster rather than a producer of energy. This EROI analysis brought into focus the need to use net energy analysis in tax and incentive policy. When Congress provided a fifty four cent per gallon incentive corn-ethanol production became profitable, although producing ethanol from corn actually wasted energy because of the negative energy balance of the process. In addition, one of its unintended consequences was that the cost of corn for food worldwide increased precipitously and caused great hardship in developing countries. Until the 1900s, over 80% of all people worked in agriculture, which had an EROI sufficient to generate enough surplus energy to support a relatively complex society. It has been estimated that prior to the 18th century, the EROI of society was between 2 and 10, depending on the efficiency of farming methods and the use of slave labor. At that time the population was relatively stable and growth was quite slow. When society began to use coal in industry, the conversion efficiency was initially quite low. But around 1950, efficiency improved enormously and populations started to grow almost exponentially. Then oil entered the energy system and the EROI of society soared to an aggregated value around 40. This high EROI provided a huge surplus of energy that fueled a population and standard of living growth unprecedented in human history.

The implications of the EROI on fossil fuel and renewable energy can be demonstrated by looking at the history of oil production during the past hundred years. During the initial phases of oil production in the United States, one unit of energy input produced almost a hundred units of energy in oil from a gusher in Pennsylvania or in Texas. When the highly productive sources of oil in the U.S. became exhausted and it was necessary to drill deeper and in faraway places such as Saudi Arabia and Alaska, the EROI dropped to about 19. The vast oil resources discovered in Saudi Arabia and the United Emirates provided a temporary supply of cheap and high EROI oil. However, importing oil increased the U.S. dependence on other countries and

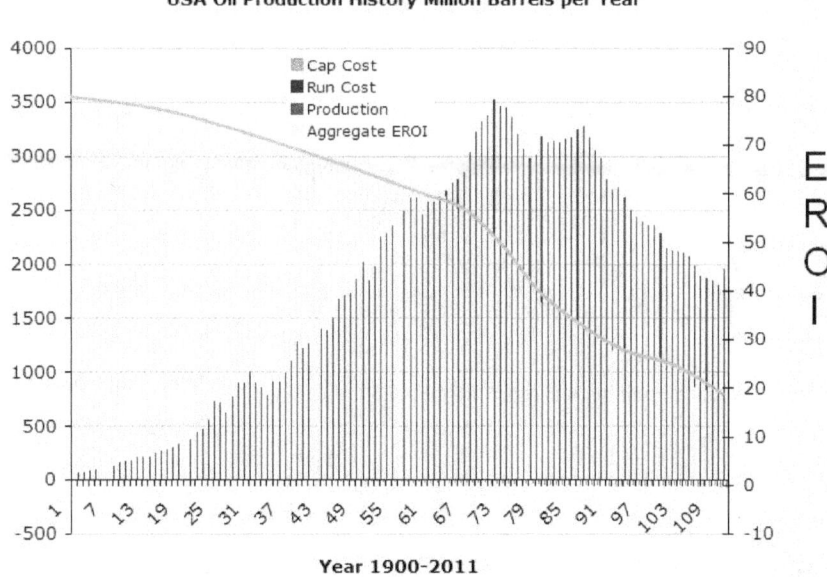

caused an ever-increasing negative trade balance for the U.S. When it became necessary to move oil production to off shore and install drilling rigs in a hostile environment, the EROI dropped further and today it stands somewhere between 3 and 6 for new installations.

In contrast to the declining EROI of oil as well as coal, the improvements in renewable energy technology manifested themselves in an increasing EROI for wind and solar thermal and electric energy. In favorable locations the EROI for wind has grown within the last fifteen years from about 8 to above 30. Similarly, the EROI of photovoltaic electricity production, especially thin film modules, has increased immensely.

The analog to the energy return on energy investment in a traditional economic analysis is the dollar return on dollar invested. Although the two are somewhat similar, their implications for future planning is different and both the traditional economic and the more modern net energy analysis should be used in planning for a gradual transition from fossil to renewable energy. There is general agreement that the unprecedented growth in human population, the improvement in the global economy and the per capita living standards of the past 150 years were made possible by the availability of cheap and plentiful fossil fuels. But that situation is changing rapidly as will now be explained. The schematic diagram below demonstrates the energy budget for society, irrespective of the energy source.

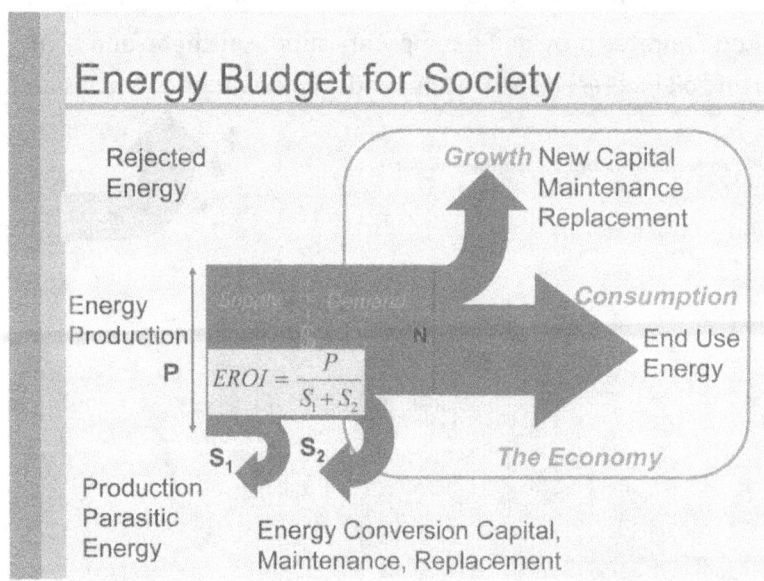

Energy Budget for Society

$$EROI = \frac{P}{S_1 + S_2}$$

I will illustrate the schematic by referring to the use of oil to power our

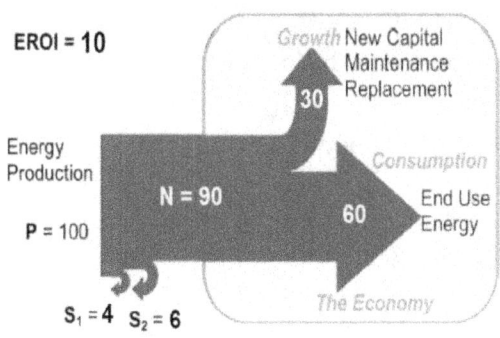

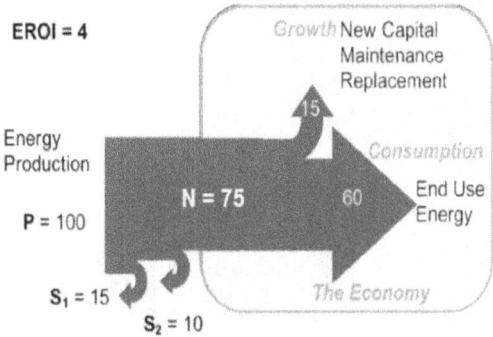

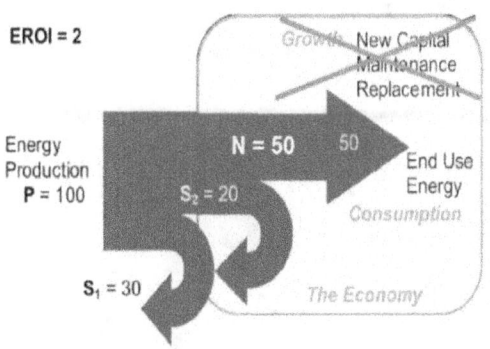

The energy budget diagrams for a society with production capacity of 100 energy units and EROI = 20 (top), EROI = 4 (middle), and EROI = 2 (bottom).

transportation system. Let us start with an oil well in Saudi Arabia. After the raw petroleum emerges from the ground, the natural gas contained in it is flared uselessly into the atmosphere as rejected energy. The liquid petroleum is then loaded on a ship or put in a pipe line and transported to a refinery. We will call the energy leaving the source during the life time of the oil well, P. In the process of transporting and refining the petroleum into useful products such as gasoline or oil, parasitic energy is lost, as shown by the arrow called S1. For the energy conversion process, an initial energy and financial capital investment is required to build the pipeline and refinery and, in addition, energy is required for its maintenance, operation and eventual decommissioning and replacement as shown by S2. The difference between P and (S1 + S2) is the net energy N entering the economy during the life of the well. Part of this net energy is used for the operations of society, such as making fertilizer for agriculture, keeping our homes warm and running automobiles and trucks. The difference between what is consumed and the net energy from the original energy source is available for growth, improved living standards, new capital equipment

and its maintenance. The EROI as shown in the diagram is the ratio between the energy production P and the sum of (S1 + S2). As the energy production process begins to require more and more energy, as for instance for the building of offshore platforms to obtain more petroleum, rather than drilling into the ground, or drilling deeper for mining and shipping coal, both S1 and S2 increase, the EROI decreases, and also the net energy available becomes less. The result is that the energy available for growth diminishes unless one reduces the energy consumption, as for example by conservation measures. The schematics below show the energy budgets for EROIs of 10, 4 and 2 respectively. Our current average EROI is about 10 while around 1950 it was about 20.

It is apparent that growth will have to be curtailed as the EROI decreases and modern society cannot function with an EROI of 3 or less. But even an EROI of 3 or 4 would severely curtail our current way of life and would leave no surplus energy for growth. The essence of this analysis can be summarized in the graph below. The ratio N/P is the net energy available for the operation of society and growth compared to the energy available from the source after parasitic losses and maintenance have been subtracted. For example, if the energy available to society were three fourth the energy from the source, the EROI would be equal to 4.

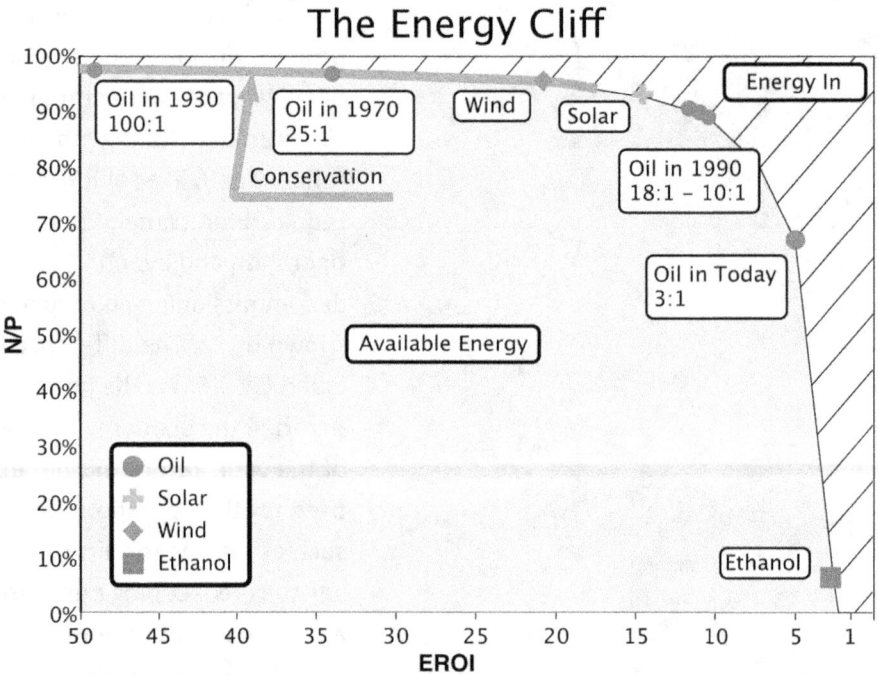

The cross-hatched area is the energy input from the source and the gray area the available energy from that input, showing the ratio N/P as a function of EROI. It can be seen that as the EROI drops below about 4 or 5, the ratio of net output to energy production input, N/P, drops precipitously. This is called the energy cliff.

Over the past ten years these concepts have been demonstrated by several people. Also Ron West and I have made EROI analyses and in one of our publications with

Cutler Cleveland demonstrated that as a result of technical improvements the EROI for a parabolic solar collector delivering heat at150 degree centigrade in Denver increased from 3 to 6 between first and third generation collector designs in about 10 years.[4]

It is unfortunate that the net energy concept is not understood by so many well meaning environmentalists who demand that governments and utilities immediately switch from fossil to renewable technologies. They are unaware that the construction of renewable energy systems will require an enormous amount of fossil energy and money input, since today 83% of our total energy use is derived from fossil sources. Engineers know that the time scale of transition will be long and will require huge financial investments that can only be obtained by enormous changes in our politics and current lifestyles. That is the reason why so many politicians pay lip service to the idea of sustainability, but are unwilling to spell out the details of the steps necessary to attain it. The most important of these is the acceptance of little or no growth and the need to conserve energy in all our activities. However, although this may lower the standard of living for some affluent members of our society, this does not imply a less satisfying life style for most of us. The UN has made a survey of the human happiness index in various countries as a function of energy consumption per capita. As shown in the graph, many countries such as France and Denmark have a very satisfying life at a much lower energy use than the U.S.

---

[4] Cleveland, C.J., Herendeen, R., Kreith, F., and West, R. Solar Parabolic collectors: successive generations of better net energy and exergy producers. ASME 1987.

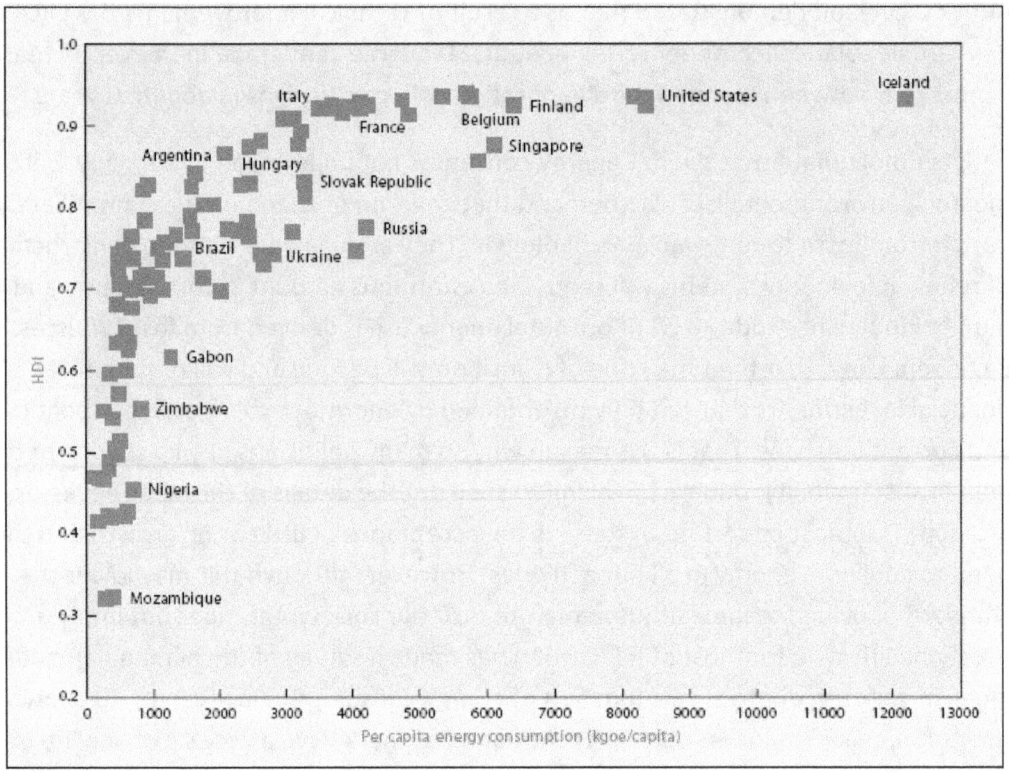

Concern about energy waste resulted in 1974 in federal legislation requiring net energy analysis of federally supported energy projects with the words "the potential for production of net energy by the proposed technology at the state of commercial application shall be analyzed and considered in evaluation proposals" [public law 93-577, sec. 5]. At the time I was introduced to net energy analysis I was unaware that it had been known to the DOE for some time. In fact, public law 93-577 required the DOE to analyze the potential net energy needs of any new energy technology before funding it and a GAO report to congress of July 26, 1982 stated that DOE has spent hundreds of millions of dollars on projects without doing this. The report further states that a reason DOE officials gave for neglecting EROI requirements is that the methodology was not feasible and rejected its utility for policy makers. That position was unreasonable, but no textbook heretofore introduced students to net energy analysis and explained how to do it. I therefore made a major effort in my latest book to include net energy analysis alongside the traditional economics and named the chapter "Economics and Net Energy Analysis."

While I was working on expanding my book for the second edition to include these topics, I had the good fortune of meeting a kindred soul, Professor Susan Krumdieck, from the University of Canterbury in New Zealand. Susan was taking a sabbatical leave at the University of Colorado and we met at a party celebrating the ascent of the highest mountain in the South Polar Region by a common friend, Professor Patrick Weidman, who had just recently retired from the University. Susan did not have an office at the University, and since mine was spacious enough to accommodate two people, I invited her to share my space. Susan accepted my invitation and it turned out that we had a very similar outlook on the energy situation, as well as on the need to plan for a transition from a fossil based system to a renewable one. Susan Krumdieck is not only an outstanding teacher, but also the National President of Engineers for Social Responsibility and CEO of the transition engineering firm, EAST Research Consultants, Ltd. This company works on developing engineering analysis and modeling tools for a transition to energy systems that greatly reduce fossil fuel demand. In fact, Susan believes that a new inter-disciplinary field called Transition Engineering, will emerge as the way our society will accommodate to a reduction in fossil fuel use and its social and environmental impacts. As an outgrowth of the many discussions Susan and I had, we jointly developed a short course entitled Sustainable Energy Systems for the International Mechanical Engineering Congress held in November 2011 in Denver. It was oversubscribed and attended by engineers from six different countries.

My book on sustainable energy systems was making good progress at the time I met Susan, but still needed a chapter on transitioning from fossil to sustainable energy systems. In view of Susan's interest in this very topic, I invited her to contribute a chapter on this topic for the new book and she accepted my invitation. The chapter she submitted two days before the manuscript was due was somewhat different from what I expected, since it did not provide a pathway for the transition. Susan is an innovative, out-of-the-box thinking engineer and, although the chapter does not give specific guidance, it does outline an overall approach to how complex systems can adapt for survival with guidance for future planning in a setting where fossil fuels will become less available and more expensive. I hope her chapter will be a useful addition to the more traditional 12 chapters which form the bulk of the book.

When the publishers asked me to suggest a cover for the second edition of the book, I recalled a famous photograph, which the nature photographer Galen Rowell called, "The most influential environmental photograph ever taken." It is a photograph taken by the astronaut William Anders during the Apollo 8 mission on Christmas Eve, December 24, 1968 from a lunar orbit. It shows our Earth rising above the moon. I thought that this photo portrays the essential message of the book as enunciated by President Jimmy Carter in 1977: "With the exception of preventing war, this [the energy crisis] is the greatest challenge our country will face during our lifetimes."

CHAPTER 41—A PATH FOR TRANSITION FROM FOSSIL TO SUSTAINABLE ENERGY, 2014

"I'd put my money on solar energy…I hope we don't have to wait till oil and coal have run out before we tackle that"
Nobel Prize Winner Albert Einstein, 1931

For several years I have served on the ASME Energy Committee which consists of professional engineers who were hoping to develop a national energy policy which could be endorsed by the members of the parent organization and then transmitted to the White House. However, over the years the members of the committee who believed in sustainable renewable energy became disillusioned with the narrow-minded outlook of the pro-nuclear majority, which is intent on promoting nuclear energy in favor of any other options. As a result, most of the proponents of solar energy resigned from the committee and it became virtually a spokes-organ for nuclear power. In fact, one of the most vocal members on the committee is also the President of an organization entitled "Go Nuclear, Inc" and another called "Environmentalist for Nuclear Energy – United States".

In November 2013, I received an e-mail from the committee informing the members that on November 17th, CNN would present a film entitled "Pandora's Promise – How Can We Power Modern Civilization without Destroying It? ". The film had previously won the 2013 Sundance Film Award for best documentary addressing environmental challenges and I expected a well-balanced exposition of the challenges related to global warming and fossil fuel exhaustion. To my dismay, however, the film turned out to be a blatant pro-nuclear, anti-renewable presentation that contained many technical errors and unsubstantiated claims. The film was seen by many hundreds of thousands of viewers when it was presented by CNN and since then it has become part of the Netflix library and has been released for showing in movie theatres all over the world.

I had previously been invited to give a graduate seminar at the University of Colorado on December 12th, three days before my 91st birthday. When I received the invitation I had intended to give a summary of my recent book, but after seeing the CNN film, I realized that the seminar gave me an opportunity to compare the only two long-term energy options for the future: nuclear or solar. Therefore, I decided to confront these two options in my seminar head-on and clearly delineate what each of them had to offer and what were their respective deficiencies. In doing so, I was hoping that I would be able to provide members of the audience with sufficient information to counter the opinion of those viewers of the film who may have been persuaded by the actors that nuclear energy is the only technology available to overcome our dependence on fossil fuels. Since the CNN movie was not released until December 10th on iTunes, I had merely one day to prepare my seminar and integrate excerpts from the movies that I wanted to rebut. In the following, I am relying on the skeleton of my presentation, but want to explain in more detail my reasoning for believing that at this point in time the nuclear option is both socially and financially unacceptable and that renewable sources are a better option for the development of a safe, socially acceptable and fiscally responsible technology than nuclear power.

The award-winning film begins by showing the devastation of nuclear bombs in Japan and the aftermath of nuclear accidents at Chernobyl and Fukushima. It then shows part of an address by the former Prime Minister of England, Lady Margret Thatcher, who warns of the dangers of global warming and states that" continued inaction to combat climate change is no longer an option," but does not explain what should be done. Then the movie claims that the nuclear accidents and the atomic bombs were responsible for an irrational public fear of radiation, and that this fear is completely unfounded when peaceful nuclear power plants are used to generate electric power. The pro-nuclear spokespeople in the film minimize potential danger from nuclear power plants or disregard its existence. The movie claims that current nuclear technology, such as was used in the nuclear U-boat, was outdated and that nuclear power will be totally safe when a new type of reactor, the Fast Liquid Metal Nuclear Breeder, becomes available for a future nuclear age. It also claims that since solar energy is intermittent, there would be no way that renewable technologies can ever provide base load power and a renewable energy system would leave the world in the dark when the sun is not shining. It further claims that the new type of reactor would be a sustainable and safe technology. Because it can reprocess the nuclear material it would be completely safe from terrorist threats and would not require a depository for the high-level nuclear waste which is part of the challenge for current nuclear technology.

When I watched the movie, it struck me that I had followed exactly the opposite path to that of the spokespeople in the movie who claim to have once been anti-nuclear activists, but now were strong supporters of nuclear power. In 1945, I had been one of the young Turks who had given talks about the peaceful potential of nuclear power under the guidance of J. Robert Oppenheimer. I had studied nuclear energy at Cornell and then given courses at the University of Colorado about the design of nuclear power plants. Preparing the seminar forced me to revisit the path I had taken from a supporter of nuclear power to one who, as a professional engineer, no longer considers nuclear energy the best alternative for a socially and environmentally acceptable technology in the U.S. There are countries such as Japan or Finland who lack sufficient renewable resources for their future needs and have no alternative but to turn to nuclear energy, but this is not the case for the U.S.

I asked myself, what have been the events and experiences for this turnabout? I think that there are at least three events that played a part. The first was my experience with the Atomic Energy Commission's effort to use nuclear bombs to stimulate natural gas in Colorado on a large scale as described earlier in the book. The second was the dismissive reaction of conservative political organizations to the three major nuclear accidents. These organizations and their adherents say that the Three Mile Island accident was contained and did not kill anyone; the Chernobyl disaster was the result of faulty Soviet design; and the Fukushima Daiichi meltdown and nuclear fallout were the result of a tsunami, not the fault of nuclear technology. The third event was my slow recognition that the nuclear establishment was not forthcoming with the facts regarding the continuously increasing cost of nuclear power, largely due to the need to provide more and more safety measures and the inability of the U.S. government to find a long-term depository for high-level nuclear waste anywhere in the country. This lack of openness was recently brought into the open once again when Turner Radio Network published official copies of an agreement between the International Atomic Energy Agency (IAEA) and Fukui Prefectural Government to conceal information about the damage, danger(s) and health effects from the Fukushima disaster from the public (http://www.turnerradionetwork.com/news/180-mjt). I had also become distressed about the manner in which the ASME energy committee had essentially driven out members that were not pro-nuclear and had virtually become a group claiming that nuclear energy was the only way for a sustainable energy future. Having been involved in both nuclear power and renewable energy technology, I felt that I had a more balanced view and that it was my responsibility to respond to the false claims in the film. This response also gives me the opportunity to outline what I consider to be a more economical and socially responsible path to combat global warming and achieve a sustainable energy system.

There have been a number of studies of what the country might look like after a transition to a wind–water–solar (WWS) system has been achieved. This is a different and somewhat easier task than asking how this transition can be made with minimal environmental impact, since at this point 83% of our energy use still comes from fossil fuels. One of the most informative and detailed study of what a steady-state sustainable energy system in the United States would look like was conducted at

Stanford University by Mark Jacobson, Guillaume Basouin, Zack Bauer, et al.[5] and key results are shown in the figure below. This study is particularly important because it addresses the assertion by the pro-nuclear establishment that a WWS system would simply require too much space to be accommodated without major social disruptions. The figure shows, however, that the amount of land area required for a completely sustainable energy system supplying the country with all of its energy needs is quite modest and should not interfere with other functions such as agriculture, housing, and transportation. However, the key challenge that still must be addressed in more engineering detail is how a transition from the current fossil to a future sustainable energy system can be achieved with minimal environmental impact. There has been enormous growth in photovoltaic and wind systems over the past decade. This is an immense feat, especially for the wind industry, but an undeniable fact that still faces the transition plan is that with all of the accomplishments claimed by solar enthusiasts, the total energy supplied by PV is still less than 1%, and for wind less than 3%. Thus, the challenge for a total transition is enormous and has to be carefully planned.

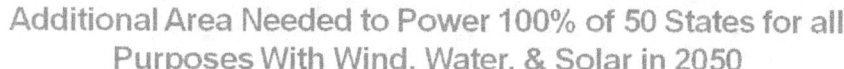

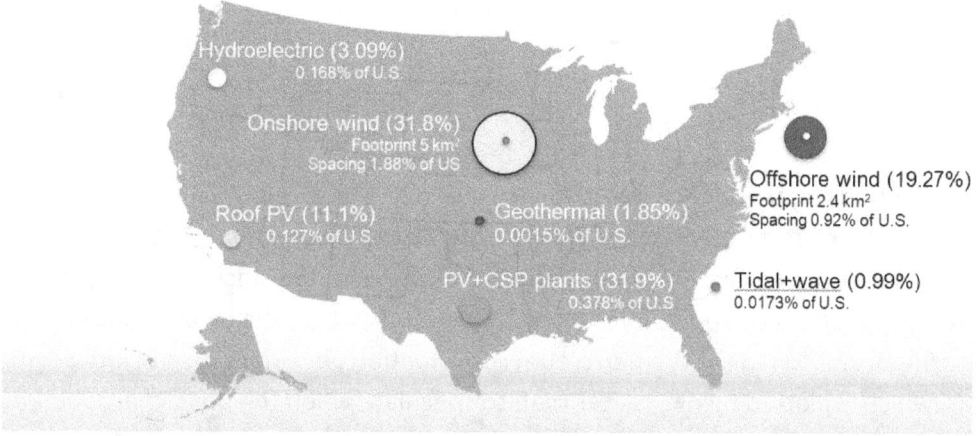

Additional Area Needed to Power 100% of 50 States for all Purposes With Wind, Water, & Solar in 2050

A recent survey by the Sandia National Laboratories provides an overview of the way the U.S. uses energy and the sources from which this energy is obtained. It is apparent from this so-called spaghetti diagram that the largest percentage of energy use is for electricity (38%), followed by transportation (27 %). The total energy used in the year 2014 was almost 100 Quads where one quad is 10 to the power of 12 (1,000,000,000,000) thousand Btus. This is an enormously large number that is hard to comprehend for the average person. The distance between the Earth and the Sun is only 92 million miles, thus to reach 10^{12} you would have to commute back and forth

---

[5] Jacobson, M.Z., Bazouin, G., Bauer, Z. et al. "100% Wind, Water, Sunlight (WWS) All-Sector Energy Plans for the 50 United States," Draft. www.stanford.edu/group/efmh/jacobson/Articles/I/WWS-50-USState-plans.html

more than 5,000 times.

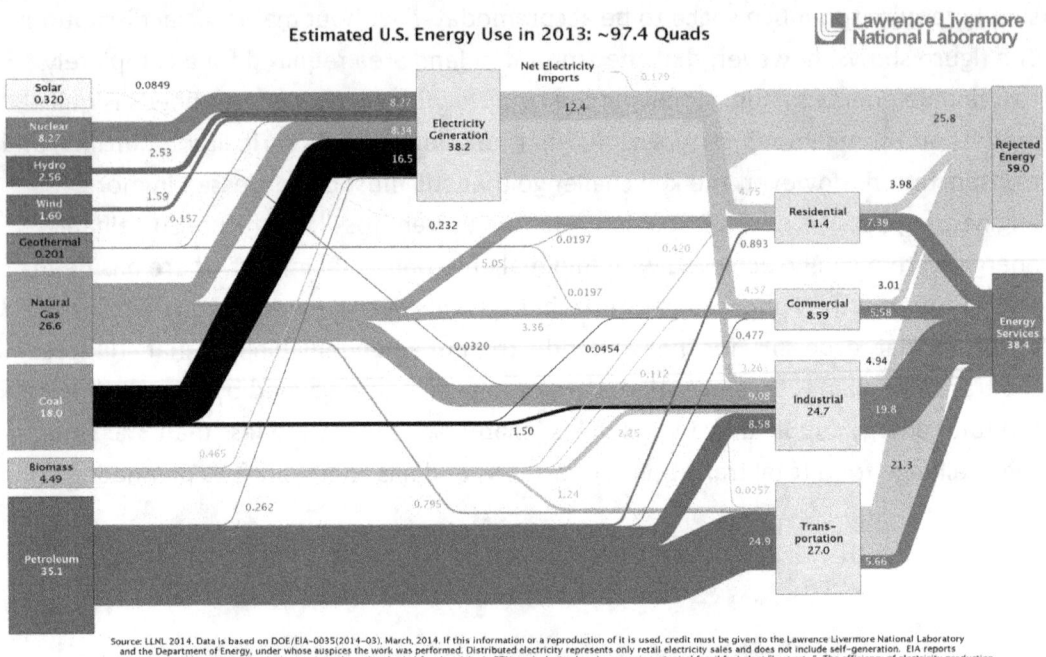

Estimated U.S. Energy Use in 2013: ~97.4 Quads

Of the two major energy users, transportation has received the most attention because a peak in oil production is widely expected to occur soon. . There are many opinions as to when the peak in oil production will occur, but since the ever-growing global population is so dependent upon liquid fuels for its transportation system, it is more important to look at the petroleum consumption per capita on a worldwide basis, which is plotted below, rather than to speculate when the peak in production will occur. It is apparent that the peak in petroleum consumption per capita passed several years ago and there will never be a day with as much gasoline per capita in the future again.

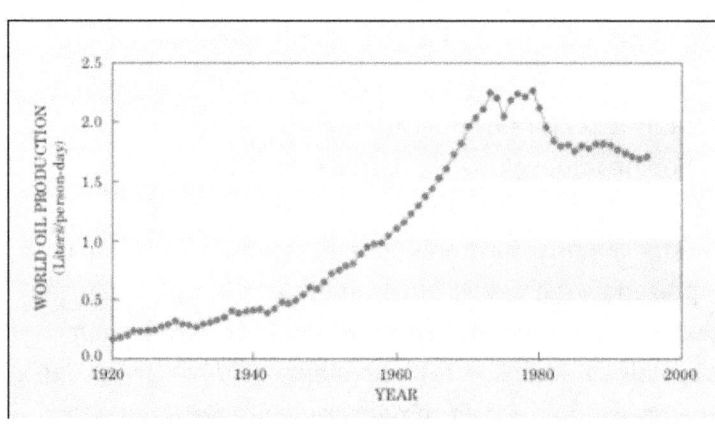

As mentioned previously, Ron West has collaborated with me in several studies and publications that have shown that changing from gasoline SI engine-driven cars to plug-in-hybrid automobiles is a feasible transition path that would decrease the oil needed for transportation substantially and also reduce CO_2 generation. The bar graph shows the miles per gallon of liquid fuel or gasoline for the current SI fleet and one that uses HEV and PHEV technology[6]. The number after PHEV denotes the miles the car can travel on battery power alone. Our estimates have recently been verified by others. For example, BMW claims that its 2014 PHEV model can get 80 miles per gallon of gasoline (*Solar Today*).The PHEV path has also been supported by the US National Research Council in a report entitled "Transition to Alternative Transportation Technologies –Plug-in-Hybrid Electric Vehicles"

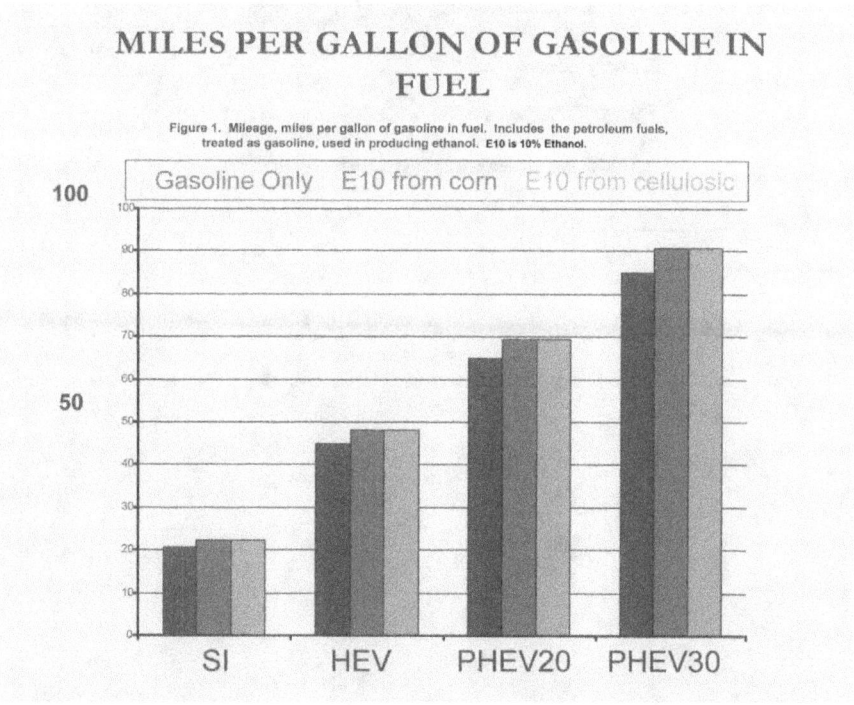

MILES PER GALLON OF GASOLINE IN FUEL

Figure 1. Mileage, miles per gallon of gasoline in fuel. Includes the petroleum fuels, treated as gasoline, used in producing ethanol. E10 is 10% Ethanol.

Gasoline Only   E10 from corn   E10 from cellulosic

SI   HEV   PHEV20   PHEV30

---

[6] West, R.E. & Kreith, F. "A Scenario for a Secure Transportation System Based on Fuel from Biomass." *Journal of Solar Energy Engineering*, May 2008, vol. 130.

Today virtually all car companies are producing and selling plug-in-hybrid electric cars that can travel as far as 30 to 45 miles on battery power. The batteries of these cars can be charged overnight or when they are parked after commuting at work without building a new electric infrastructure. Some car manufacturers, such as Tesla, are betting on all electric cars hoping to develop new batteries that can be charged quickly and travel long distances. Only the future will tell whether or not this expectation will come true. But for the near term there is a reasonable transition path with available technology supported by the public, government and industry that can replace a large fraction of oil consumption. If this electric technology were combined with bio-fuels, transition to a sustainable transportation system could be achieved in a reasonable time. To pay for the transition the government could put a sizable tax on gasoline, similar to those in European countries, and use this money to allocate credits or subsidies for battery powered plug-in-hybrid electric vehicles in a smooth and timely transition from current SI driven technology.

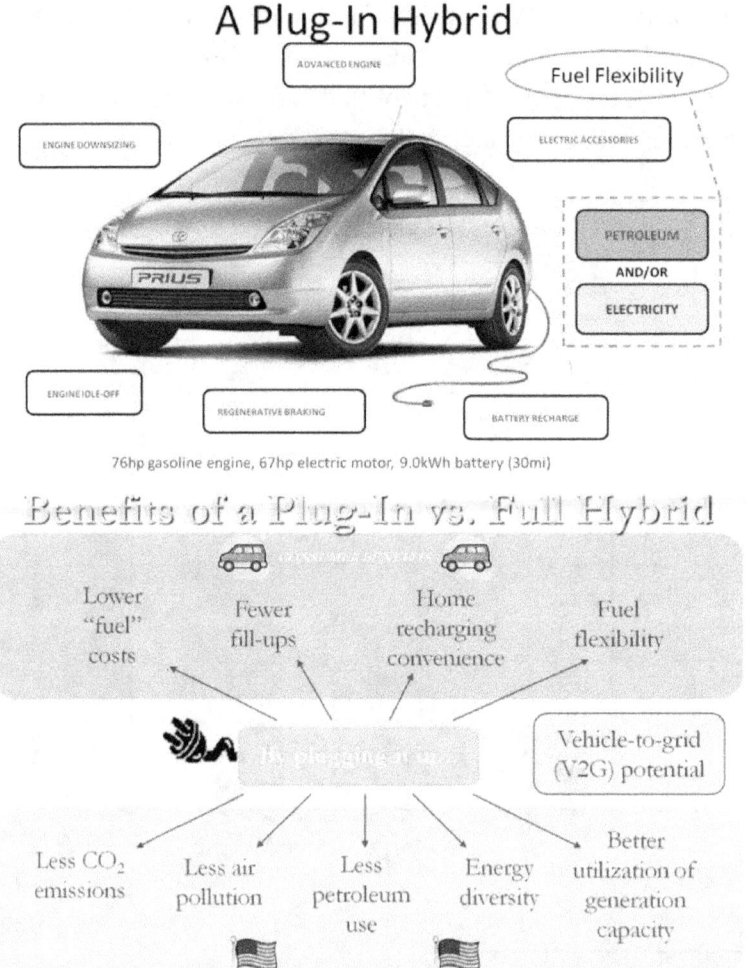

A Plug-In Hybrid

ADVANCED ENGINE

Fuel Flexibility

ENGINE DOWNSIZING

ELECTRIC ACCESSORIES

PETROLEUM

AND/OR

ELECTRICITY

PRIUS

ENGINE IDLE-OFF

REGENERATIVE BRAKING

BATTERY RECHARGE

76hp gasoline engine, 67hp electric motor, 9.0kWh battery (30mi)

Benefits of a Plug-In vs. Full Hybrid

Lower "fuel" costs

Fewer fill-ups

Home recharging convenience

Fuel flexibility

Vehicle-to-grid (V2G) potential

Less CO$_2$ emissions

Less air pollution

Less petroleum use

Energy diversity

Better utilization of generation capacity

The situation is very different, however, for transitioning from fossil fuels to sustainable sources in the electric power sector. When I started this book there were two major questions I could not answer satisfactorily. The first was the assertion of pro-nuclear engineers that renewable sources could not be the answer to a sustainable electric energy system because the sun and wind are intermittent, and when the sun is not shining and the wind is not blowing, our houses would be freezing and industry would come to a stop. Thus, base-load nuclear power to produce electricity is needed. The other question was that, although we are aware that a transition is inevitable, the path for a planned and peaceful transition from fossil to renewable energy systems was not clear because currently 83% of our energy comes from fossil fuel and all of our energy production is used to sustain our current way of life. Where would the excess energy come from that is needed to build a new sustainable electric energy system without destroying our current social and energy structure?

An answer to the first question came from a study conducted by two scientists at the National Renewable Laboratory, Dr.Walter Short, an American economist, and Dr.Victor Diakov, a Russian born chemical engineer.[7] They asked the question: if we could place solar PV panels and wind turbines in the most appropriate manner across the United States and a robust transmission system for electric power were available, what percentage of our total electricity need could be provided from wind and sun? Short and Diakov calculated that wind and solar alone could provide about 50% of our electricity needs. But if, in addition, 40 Megawatts in available pumped storage capacity were used, 83% of our electricity needs could be met. More than 20 Megawatts of hydroelectric storage are already built and pumped storage is a well developed technology which uses excess energy when available to raise water to some elevated location and then uses a water turbine at the bottom to provide electric power by releasing the potential energy in the water as needed. This is quite similar to the ancient water wheel and is currently used by utilities to provide extra power during high demand periods, such as air conditioning during a hot summer evening. The map below shows the optimal WWS locations, based on many years of weather data accumulated by the National Renewable Energy Laboratory, to supply 83% of the electricity needed. This is comparable to the utilization factor of nuclear power plants that is somewhere between 85-89% according to recent data. If we were to develop a nuclear energy system, we would need a large number of new nuclear power plants distributed across the country.

---

[7] W. Short and V. Diakov. Renewable Energy Load Matching for Continental U.S. ASME 2011 International Mechanical Engineering Congress & Exposition.

Wind and PV can supply 83% of U.S. electric power needs as shown below without water consumption.

Number of nuclear power plants needed to provide 25% of energy needs as electric power. Each dot is for two 1000 MW reactors.
(Courtesy of Prof. Gary Vliet)

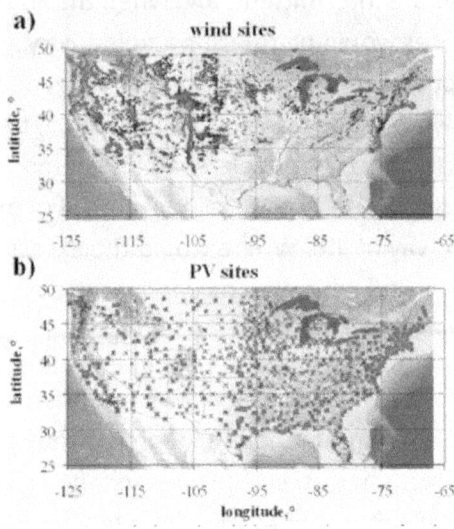

a) wind sites

b) PV sites

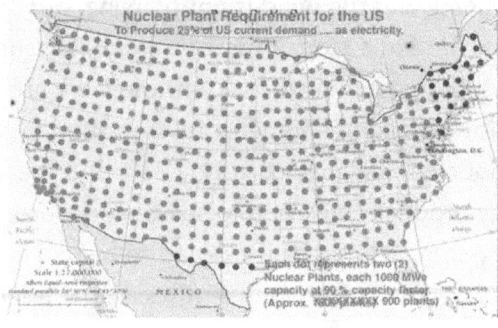

A map of nuclear power plants provided by Professor Gary Vliet of the University of Texas shows the number of nuclear power plants required to provide 25% of the country's energy need as electricity. Each of the dots on this map is equivalent to two 1,000 MW nuclear plant, approximately equal to the two plants that were involved in the Three Mile Island accident in Duaphin County, Pennsylvania. When such large numbers of nuclear power plants have to be sited, it is inevitable that some of them are going to be near fault lines susceptible to earthquakes and the cooling water requirement necessary to operate these plants may create stresses beyond local capacity. In contrast to the nuclear path, wind and PV hardly require any water and the water used for pumped storage can be largely re-circulated. Thus, I believe as an engineer that the Wind, Water, Sun approach [WWS] is the safer and environmentally preferable way to achieve sustainability.

The remaining 17% needed to meet peak electricity requirements could be provided in a number of ways. A recent DOE Report (http://nhaap.ornl.gov/nsd) provided an updated hydropower assessment in the US. There is still a large potential in new hydropower development in rivers and streams. The largest potential capacities are in the Pacific Northwest (25 megawatts) and in Arkansas (16 megawatts) and Missouri (12 megawatts). In addition one could use electric power stored in the batteries of plug-in hybrid electric vehicles, continued use of the best existing nuclear plants that have already been built and paid for or by small natural gas turbines. And of course there is always use reduction by conservation.

A good deal of energy is also required as heat at intermediate and high

temperature for industry. Solar thermal energy from concentrated sunlight in sun-tracking solar trough collectors is a technology well suited to meet the needs of industrial process heat which is a large portion of the energy portfolio. This technology is fully developed and thermal storage of energy is available for any thermal solar collector technology. A solar trough concentrator at a test site near Boulder shows a fully developed single axis tracking solar concentrator.[8]

Either the nuclear or the renewable approach would require the investment of an enormous amount of capital. The latest estimate for one of the two nuclear plants approved in 2013 to be built in Georgia was about $8,000 per kW. Thus to build approximately 1,000 plants with 1,000 MW capacity each would cost on the order of 8 trillion dollars. The construction of a similar WWS system with current wind construction costs at 2,100$-/kW and PV at 2,790$-/kW and capacity factors of 35% and 25%, respectively, would be on the order of 6.3 trillion dollars. This compares to the Gross National Product for the year 2013 of approximately 17 trillion dollars. However, as shown below, the cost of wind and PV power is decreasing, while as mentioned previously, the cost of nuclear power according to data from the U.S. and France has been increasing enormously, largely due to additional safety requirements mandated to avoid accidents.[9]

---

[8] Picture used with permission of Casey Cass, Photograph for the University of Colorado, Boulder.
[9] U.S. Department of Energy, "2011 Wind Technologies Market Report".
U.S. Department of Energy, "Photovoltaic (PV) Pricing Trends: Historical, Recent, and Near-Term Projections".

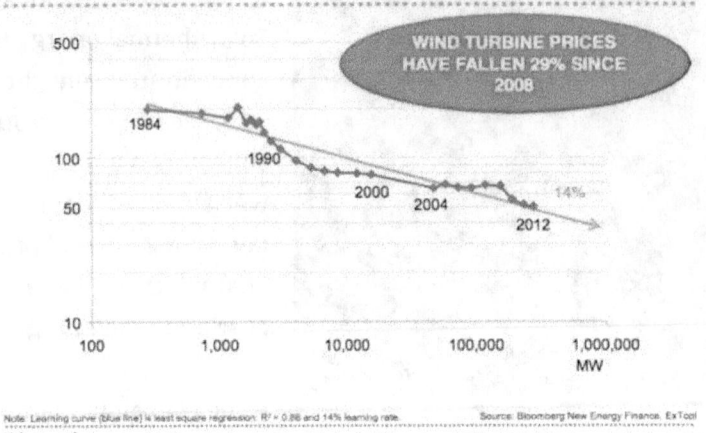

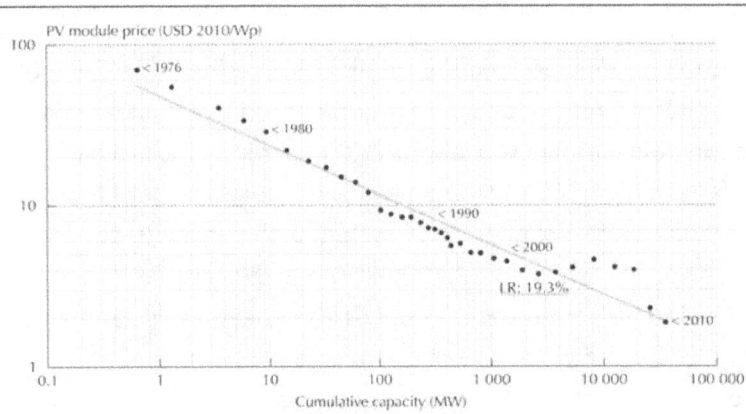

Figure 6.3 **The PV learning curve**

Source: Breyer and Gerlach, 2010.

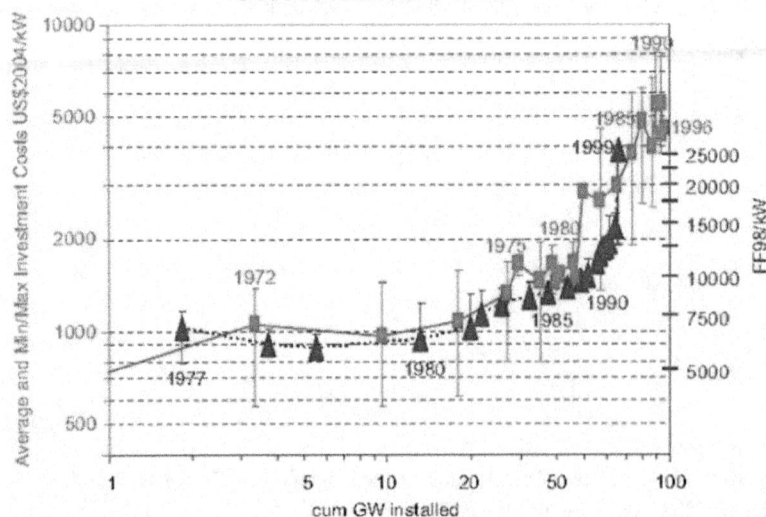

Cost of Nuclear Power

Source: Joe Romm,2011.

My other concern about how to plan for a smooth transition to a sustainable energy system received an unexpected, but welcome solution by an engineering development called directional drilling, which is the practice of drilling non-vertical holes from a single vertical shaft deep down in the earth. A schematic diagram of this practice shows that one can drill vertically to a depth far below the water table and then turn the drill bit so that it can continue drilling targets located laterally far from the surface location of the vertical drill bore hole.

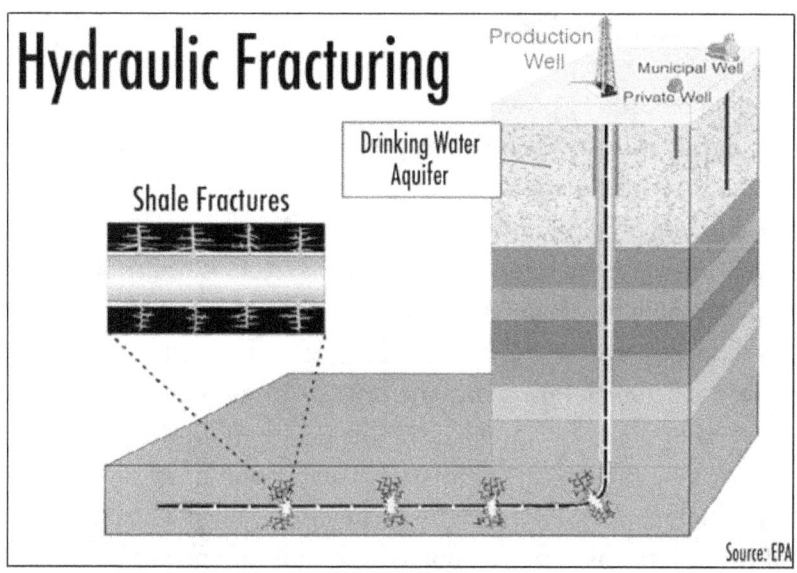

It has been shown that this technique can reach horizontally up to six miles away from a vertical hole at a depth between 5,200 - 8,500 feet. The technology was developed further with a camera to direct the drilling. This new drilling technique has opened a window of opportunity to obtain natural gas and oil by a technology known as fracking. For a long time, the gas contained in tight shale formation 5,000-8,000 feet underground, was not economically accessible. But with the new drilling technique, a single bored hole can explore a vast area and oil companies began to use this technique to inject water mixed with sand and some chemicals into the tight formation, thereby loosening them to allow the gas to escape. Fracking had been previously used for at least 50 years by Halliburton and other large companies, but as long as only vertical drilling was technically feasible, it was not a financially successful technique, as demonstrated by the effort to use nuclear devices for the fracking operation.

Given this development, I began to view the newly available natural gas from shale and the transition to renewable energy as a symbiotic relationship. We have already seen that the cost of renewables is decreasing and the availability of relatively inexpensive natural gas provides an opportunity to build sustainable systems without having to use the existing energy supply sources. Energy production with natural gas produces less than half the carbon dioxide emissions per unit of power than coal. Thus, gas can be used as a transition fuel to a renewable, low carbon world. In order to make the natural gas from fracking a vehicle to a sustainable energy system in the U.S. it is necessary to abstain from exporting the excess natural gas and use it to build our own sustainable system. This approach to transition will require strong leadership and public support which has heretofore been lacking.

There are many outstanding engineers and physicists who believe that the solution to the world's energy problem will be the development of a fusion reactor. This technology would essentially create the equivalent of a small star at such high temperatures that it could be used to provide a virtually unlimited source of power. Some of the believers in the future of fusion energy have asked me why I do not share their vision for building a sustainable energy system. My response to that question is that I am hopeful that someday their vision will come to fruition, but the time available to make the transition from a fossil to a renewable energy system is too short to wait for the realization of a workable fusion power plant. I am aware of the International Thermal Nuclear Experimental Reactor (ITER) which is under construction in the south of France. Many countries, including Russia and the United States, are contributing to this audacious plan, but even the most optimistic of the engineers working on the project realize that it may take many years before even an experimental reactor can be built and tested. In my opinion, it would therefore be unwise to build a plan for a transition from fossil fuels on the assumption that a fusion nuclear power plant will be available in time. An excellent history of the attempt to build a fusion thermal nuclear reactor, often called a Tokamak, can be found in the March 2014 issue of The New Yorker entitled "A Star in a Bottle".

The production of shale gas through fracking is the most important development in the U.S. energy sector in generations. But the technology has faced public opposition since its use became widespread a few years ago. Public fears include contamination of drinking water supplies from some chemicals used and air pollution from methane escaping into the atmosphere. These fears are not unfounded under current regulations because the Energy Act of 2005 exempted fracking from EPA regulations under the Safe Drinking Water Act, except for diesel fuels. A core element of this act is setting requirements for proper well citing, construction, and operation to minimize risks to underground sources of drinking water. Drilling companies claim that the chemicals are safe, but have used this exemption to consistently refuse to provide a list of the chemical additives used in fracking fluids. To convince the public that fracking can be done safely this lack of openness and evasion must stop because the technology exists to mitigate these dangers as explained in a guest opinion I wrote for local newspapers in 2013.

Can fracking be safe?

By Frank Kreith

Boulder County's extension of the temporary moratorium on processing new oil and gas drilling applications (Daily Camera, Jan. 25) has essentially put on hold any new oil and gas drilling that uses the hydraulic fracturing process called "fracking." This is an opportunity to examine the broad environmental aspects of fracking from an engineering perspective.

Until recently, underground supplies of gas provided only 1 percent of the total natural gas consumption. As a result of developments in hydraulic fracturing combined with advances in horizontal drilling, today natural gas from tight formations provides about 30 percent of our total consumption. This could be good news for the environment, because burning natural gas emits less carbon dioxide than combustion of coal or building a nuclear power plant and extracting uranium to operate it. But fracking, like any energy production technology comes with some risks and public fears have recently been expressed about contamination of drinking water supplies from chemicals used in fracking and air pollution from methane. There is no question that public safety should be a primary concern in the development of any energy source. The question is whether public fears about natural gas from fracking are sound and whether measures can be taken to avoid deleterious contamination from fracking activities.

A lethal fracking experiment was conducted in northwest Colorado in 1969, when the Atomic Energy Commission (AEC) exploded an atomic device, about twice the power of the Hiroshima bomb, approximately 150 feet underground in the Piceance Basin in an effort to loosen the rock formations and facilitate the collection of natural gas. This was followed by a second explosion of three 30 kiloton nuclear bombs near Rio Blanco just three years later. At that point, the citizens of Colorado passed a referendum in 1974 that put in the constitution a requirement that no nuclear

fracking can be conducted in Colorado without the specific consent of the people. This action was necessary before because gas from nuclear fracking was a radioactive health threat, but the gas obtained with current fracking technology is no different than the natural gas obtained with ordinary drilling.

Like any energy technology, fracking has to be carefully controlled. The question is whether or not current fracking technology can be regulated to assure the safety of the people and the environment.

A main shortcoming in assuring that fracking does not contaminate groundwater is the "Halliburton Loophole." When Vice President Dick Cheney, a former executive of Halliburton, chaired President George W. Bush's Energy Policy Task Force for the Federal Energy Policy Act of 2005, he championed a provision that exempts fracking from federal oversight for gas drilling and extraction from requirements in the Underground Injection Control (UIC) program under the Safe Drinking Water Act of 1974. This provision delegates regulation of fracking to individual states and without the safeguard under the UIC program, drilling companies can simply claim that the chemicals used in the fracking process are safe, but at the same time, refuse to provide a list of the chemicals additives used.

An important issue for the future is the amount of new gas made accessible by fracking and the length of time these new sources will last. In 2007, the National Petroleum Council estimated that the natural gas reserves were 1,450 trillion cubic feet (tcf), which at the current rate of consumption would last 60 years. Four years later, the U.S. Energy Information Agency increased this estimate by including fracking, to 2,550 tcf, which would last more than 100 years. But, if the annual consumption were to continue its historic growth rate of 2 percent, the projected lifetime would decrease to 64 years, and if natural gas were to be used for transportation, the lifetime would shrink to 33 years.

Any energy production technology has some environmental impact. Even renewable technologies such as wind and photovoltaics require a large fossil energy input in materials and energy for building the systems. Of all the fossil fuels, natural gas has the least environmental impact, and that factor must be considered when weighing the pros and cons of producing it by fracking technology.

To minimize adverse effects of fracking, several steps are necessary. First, Congress needs to close the Halliburton Loophole and mandate public disclosure of all chemicals used in fracking. Secondly,

regular inspections of the concrete casings in the well bores have to be mandated in order to prevent groundwater contamination due to failure in these casings. Third, fluid byproducts of fracking must be stored in closed tanks. Fourth, gas companies should be required to inject tracers with any potentially harmful chemical in the fracking fluid to ascertain whether any of the fluids end up in the water supply and stop the fracking if it does. Finally, municipalities should test aquifers and drinking water in wells for chemicals before drilling begins, and continue this testing as long as extraction goes on. Any changes in the quality of groundwater would automatically require discontinuance of fracking in the vicinity. We should also promote recycling of water used in fracking fluids and consider the aesthetics of the drilling equipment in the permit process. Additional safeguards may become necessary as experience with this technology is accumulated, but there is no reason why fracking cannot be made as safe or safer than any other energy extraction method and provide us with a transitional energy source that can help build a truly sustainable energy system.

Frank Kreith is professor emeritus of engineering at the University of Colorado, and author of "Principles of Sustainable Energy."

I believe that the natural gas newly available from shale and the need for a transition to a renewable energy system can form a symbiotic relationship. We have already seen that the cost of renewable energy is decreasing and the availability of relatively inexpensive natural gas provides an opportunity to build sustainable systems without having to use the existing energy sources of coal and nuclear. But I realized that in order to demonstrate the viability of a symbiotic relation between fracking and sustainability, I would have to provide a quantitative analysis. For this purpose, I received a small grant from the Retired Faculty Association and advertised for a research assistant. I received more than a dozen applications, but one of them stood out. It came from Kangqian Wu, a Chinese mechanical engineering graduate student who was in the United States without any government support. I contacted the young man and he told me that he came from an area in China that had a great deal of shale and he was personally interested in learning more about the potential of obtaining natural gas from shale formations. I arranged for an interview with Kangqian in my office the next day. The young man was slender, of average height, dark hair and had a gold-winning smile with a friendly air of self-confidence. We immediately took a liking to each other. Kangqian agreed to work under my guidance and he wrote a proposal for his thesis entitled "A Symbiotic Relation Between Fracked Natural Gas and Sustainability."

Kangqian and I, with the help of my friend Ron West, developed an analytic framework for determining the amount of energy needed to build a WWS system and the amount of energy available from the excess natural gas production as predicted by the Energy Information Administration (EIA). Our calculations showed that the amount of energy that will be available, according to the EIA estimates, will be more than adequate to build a sustainable WWS electric power system to supply 83% of the U.S. electric power in about 15 years. However, under that scenario, at the end of 15 years when all of the electric power would be supplied by wind and solar panels, suddenly there would be no market for the existing production facilities because all the energy needs would be satisfied by wind and PV. We therefore modified the model to produce the wind and solar panels required at a slower pace of construction and showed that within less than 30 years a sustainable WWS energy system could be built without the production facility experiencing a sudden drop in demand. The demand would increase again when the first group of PV panels and wind turbines would reach the end of their life and would have to be replaced. But from this time on the energy for building the replacement wind turbines and PV panels would come from renewable sources. In this manner a truly sustainable and renewable energy system would emerge.

Our analysis also showed that the environmental impact during the 30 year transition period to a WWS system would be considerably less than if the current energy supply system would continue unchanged. The amount of primary energy used during the 30 year construction period for the WWS system would shrink from 112 million gigawatts if the current system were to continue, to 10.8 million gigawatts. Thus the WWS construction would result in a net reduction of primary energy use of about 100 million gigawatts. This is an enormous savings that would also result in a reduction in CO2 emissions of more than 30 billion tones of CO2 as shown below. This reduction in CO2 generation by the United States, which is currently the largest CO2 producing country, would hopefully be an inspiration for the rest of the world to follow suit. This reduction in CO2 generation would be accomplished without sequestration of CO2 from coal burning power plants, which is a very expensive way to reduce global warming. [10]

Reduction in CO$_2$ Emissions by Constructing a WWS System

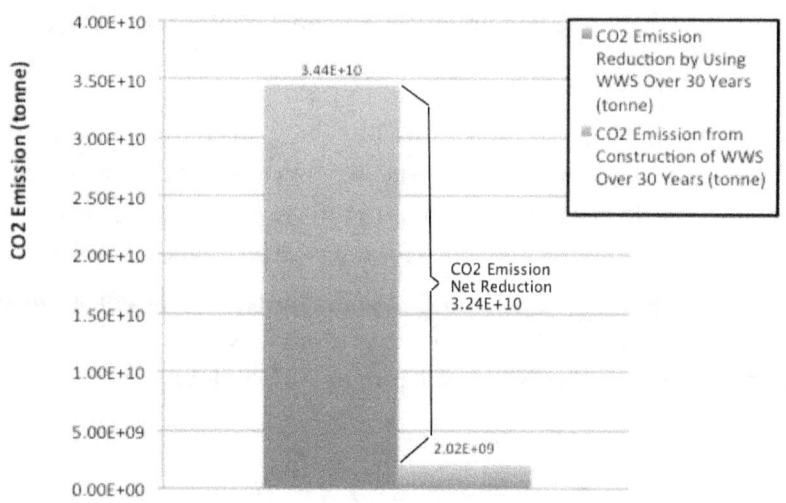

However, with any current natural gas technology, irrespective of whether fracking or conventional drilling is used, there is a significant leakage of methane into the atmosphere and methane is a very powerful global warming gas. But, whereas CO2 generation is an unavoidable consequence of the combustion of fossil fuel, methane leakage occurs largely during transmission after the gas has reached the surface and

---

[10] An Assessment of Using Natural Gas from Fracking to Build a Sustainable Energy System by Kangqian Wu. M.S. Thesis. University of Colorado at Boulder. 2013.

can be reduced substantially with currently available engineering measures, provided stringent regulation are instituted, funded and enforced by periodic inspection.

In summary, the proposed WWS transition scheme described above offers a path from fossil to sustainable energy with available technologies, minimal environmental impacts and a newly available energy source in a reasonable time. But this opportunity may not come again if the natural gas from fracking is simply burned in existing power plants or exported for political reasons.

Looking back on the challenge for energy independence given to this country by President Nixon more than 40 years ago, I keep asking myself why there has been so little progress towards an energy future that does not depend on fossil fuels. I have been actively involved in renewable energy challenges, even before President Richard Nixon proclaimed Project Energy Independence in 1973. As I have tried to show in this book, we have the technical know-how to start a transition from fossil to renewable energy now. The steps involved in planning for a sustainable energy future have been the major challenge in my past 50 years of professional activity and I believe that my personal observations of the reasons for the lack of progress in meeting the goals set more than 40 years ago by some of the most prominent energy experts of this country would be of interest to those involved today in working for a sustainable energy future. I therefore began to write this autobiography in the hope that my life experience will be of interest and benefit to the next generation during whose life time the fossil era will approach its peak.

The Nobel Prize-winning Intergovernmental Panel on Climate Change (IPCC) states in its 2014 report that "Climate change will...worsen existing global security problems such as civil wars, strife between nations and refugees...fights over resources like water and energy, hunger and extreme weather will all go into the mix to destabilize the world" (2014, Associated Press). Our choice seems clear. Either we begin to undertake a voluntary peaceful transition from fossil fuels to sustainable energy sources soon, or we face a crisis that will force a violent, involuntary change later.

EPILOGUE

In the course of my search for a meaning of my life to justify my survival from persecution of Jews in Nazi Germany, I was approached by my former high-school mate, Gerald Holzapfel, who is shown in the photo of my 1938 gymnasium class sitting at the very left side in the first row.

PROJECT SECOND WAVE

Gerald Holton
Research Professor

Gerhard Sonnert
Research Associate

Joan Laws
Administrator

358 Jefferson Laboratory
Harvard University
Cambridge, MA 02138
Tel.: 617-495-4474
Fax: 617-495-0416
Email: holton@physics.harvard.edu
sonnert@physics.harvard.edu

November 2006

Dear participant of Project Second Wave:

You may remember that a few years ago you took part in Project Second Wave by kindly filling out our survey and/or participating in an interview. Project Second Wave has been studying the group of young refugees who fled from National Socialist persecution in Central Europe to the safety of the United States in the 1930s and 1940s. While there has been a large and growing amount of autobiographical, biographical, and journalistic accounts of the fates of individual members of this extraordinary cohort, our goal was to provide, for the first time, a scholarly study of the whole group.

After several years of intense research, the project has come to a successful conclusion, as we informed you in a mailing earlier this year. Our book titled _What Happened to the Children who Fled Nazi Persecution_, presenting the major results of our research, is being released by the publisher, Palgrave Macmillan, in December 2006. (The table of contents is printed on the back of this sheet.)

As one in the group of former refugees who are the focus of the book, you may be interested in the results. Also, members of your family who may not have gone through the refugee experience themselves might like to read this record of the refugee group and their fates.

As a small token of our appreciation and gratitude for your involvement in the project, we have negotiated with the publisher the offer of a personal discount of 35% from the publisher's regular price of $69.95, specifically for the participants of Project Second Wave. If you are interested in owning a copy of the book, you can take advantage of this discount by using the discount order form at the bottom of the enclosed flyer provided for this mailing by the publisher. You may also note on the flyer the gratifying endorsements of the book.

Thanking you again for helping us in this project,

Sincerely yours,

Gerald Holton

Gerhard Sonnert

./.

Advisory Committee

Bernard Bailyn
Lotte Bailyn
W. Robert Connor
Lewis Coser (-2003)
Mary Frank Fox
Howard Gardner
Nathan Glazer
Hanna H. Gray
Inge Hoffmann
Stanley Hoffmann
Jerome Kagan
Stanley Katz
Herbert Kelman
Walter Laqueur
Kenneth Prewitt
David Riesman (-2002)
Robert Rosenthal
Neil J. Smelser
Michael Sokal
Arnold Thackray
Mary Waters
Spencer Weart

Supported by a grant from the Andrew W. Mellon Foundation, New York

Gerald immigrated directly to the United States and became a famous professor of

the history of science at Harvard University. I followed his road to success indirectly and learned that he changed his name to Gerald Holton. Sometime in 2004, I received a letter from Gerald asking me to take part in a project entitled "Second Wave".

The project studied a group of young refugees who had fled from Nazi persecution in central Europe to the United States in the 1940s. The project came to a successful conclusion in November of 2006 and Gerald sent me an announcement for a book titled, *What Happened to the Children Who Fled Nazi Persecution?,* published by Palgrave McMillan. The book summarized its finding in the form of four syndromes that a majority of the escapees experienced. The first of these syndromes was a feeling of the impermanence in any stage of the refugees' life from a sense of guilt for their good luck when so many others and more worthwhile people had perished. This guilt feeling came with a discomfort of being Jewish. The second syndrome afflicting the refugees was a lifelong deep-seated fear that motivated a drive toward socio-

economic success. This desperate striving for success was, in the mind of these refugees, a way to pay for their survival by making extraordinary contributions to society. The third syndrome Holten describes is trying to devote one's life to making the world a better place and finding personal peace in the process. Holton also mentioned a fourth theme that he called "revenge and getting even". The core of this syndrome is an abiding interest in projecting strength so as never again to be seen in a position of weakness that may precipitate persecution. Re-reading my book, I was amazed to find such similarity in these syndromes resulting from escape of a life-threatening situation by all these refugees.

Looking back on my life, I feel fortunate to have been given an opportunity to justify, at least in my own mind, my survival from the Holocaust. When I said to myself 75 years ago, "If I survive this, I do not just want to just live, but I want somehow to pay back for my survival",this was a vague goal I set for myself at a time when I had no idea what my life would be like. Today, I feel that in my own way, with the help of my wife Marion, I have been able to achieve this goal. Here we are, side by side, 62 years later in beautiful Colorado, our new home.

I am not religious, but I am proud to be a Jew. In the aftermath of the Holocaust I have lost faith in the existence of a superior being. But I feel part of the universe, especially when I stand at the shore of a lake and look at the mountains behind it or when I see a beautiful wild flower in a meadow in the sunlight. I want my children and their generation also to experience unspoiled nature. Maybe that is why solar energy seems to be such a powerful force in my life.

After escaping from the pending holocaust of Nazi Germany, I felt a deep subconscious need to accomplish "something" that would make my life count. Finding that "something" was a long and tortuous path, as this book demonstrates. But as my life progressed, I became convinced that the fossil fuel era will end soon and to survive

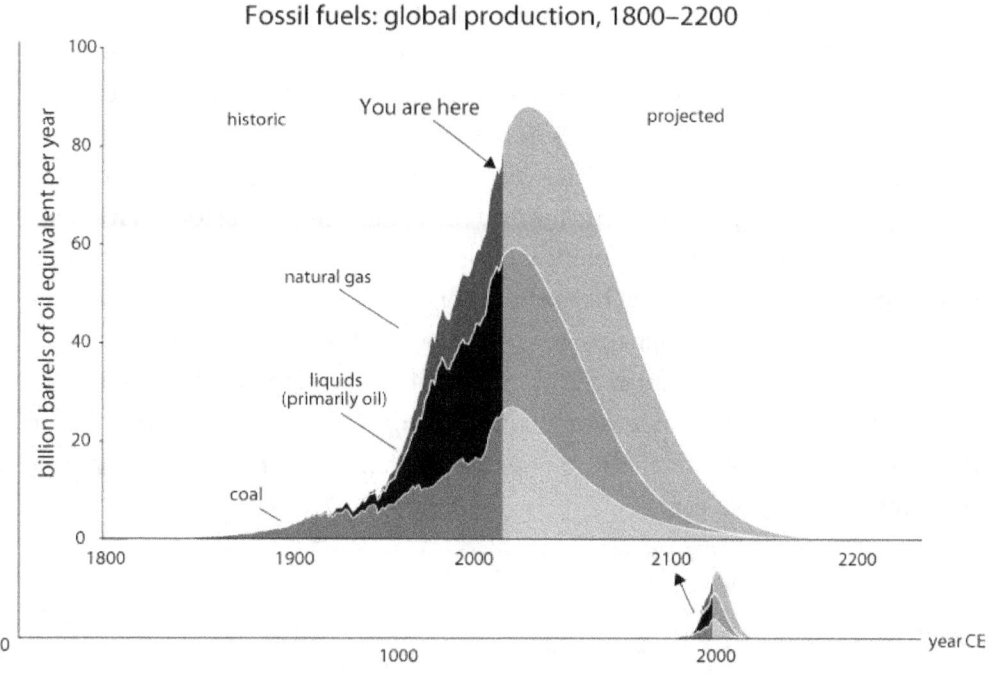

as a civilized people, we have to plan for a transition to a sustainable energy future on our planet earth. The fossil era is like a mere flicker of light in the history of evolution as shown by the world fossil fuel consumption on a geological scale. Whether or not mankind can survive after the fossil energy supply has peaked will depend on how we adapt to life with an ever diminishing supply of natural resources, especially liquid fuels. The current perspective that growth is necessary for economic health is not sustainable for the future because Energy does not obey so-called laws of economics

I have been an active part of the global energy evolution and this biography describes my experience. Renewable energy was briefly in the forefront after the Arab oil embargo in the early 1970s, but interest and support did not continue. Today political change and public awareness of the impending energy crisis are taking place, but the pace of change is far too slow to avoid a human catastrophe resulting from a shortage of food and affordable energy sources for the ever-growing number of people on the globe.

During the past 20 years, the majority of people have become aware of the need to plan for a transition away from fossil fuels—some call it decarburizing our world—in order to make our environment safe and livable. But few people, especially very few of our decision makers, seem to be aware of the enormity of the task. The use of fossil fuels imposes a large collective burden on society such as health effects from air pollution, military expenditures to protect our oil supply and adverse effects from global warming. Fortunately, awareness of the danger of over-using our global resources and the need to accelerate the transition to sustainability are happening, although at a snail's pace, and my efforts have so far not changed the pace significantly. As my friend Susan Krumdiek put it, "For better or worse, you happen to be alive at one of the most dramatic turning points in our specie's history that ranks right up there with climbing down from the trees. The only real question is what role you want to play". I believe that we have the technical knowhow to make a peaceful transition from a fossil to a sustainable energy system, but so far we lacked the willingness to make the necessary changes in our politics and lifestyle to achieve it. What the next generations will actually do with the engineering tools to which I have devoted much of my professional life no one can predict, but I am hopeful they will use them well and wish them success!

ACKNOWLEDGEMENTS

After more than 50 years in the energy field, it is virtually impossible for me to remember all of the people who have been influential in shaping my career. I will highlight the people who have had the most significant impact on my personal and professional life, but I am sure I will have forgotten many of them. All I can say is that I will do my best.

First and foremost in my personal life are all the members of my family, particularly, my wife, Marion, my late sister, Susan Culler/Kreith, my three children, Michael, Marcia and Judith, and my friend and confidante, Bev Weiler. In addition, of course, I thank all my students who have helped me to keep in touch with the ever changing social background in the real world that often escapes a person in an academic environment. My interactions with these students, particularly those in the honors program, have also helped me maintain a positive outlook towards the future. Their youthful enthusiasm has been an inspiration for me.

There is my friend from the Vienna school days, Paul Toch, who later invited me to join him in California after our escape from Europe. We shared many camping trips into the wilderness of the Sierras, including a memorable outing on skis on September 7, 1941. We did not know that the Japanese had attacked Pearl Harbor until we returned two days later. Another person who had a great influence on my life is Don Ellis. I met him first in one of the early honors courses and we developed a lifelong friendship, largely through our common love of nature. He taught me a lot about camping in the wilderness and also tried to teach me rock climbing. I did manage to make a few minor ascents with him, but after we had a near fatal accident on one of them, I decided to discontinue this beautiful, but addictive sport. I owe a debt to the late Paul Rappaport, the first director of the Solar Energy Research Institute, who shared and supported my ideals for a sustainable energy future. I owe an enormous debt to my personal friend and professional partner, Ron West, who has supported and properly critiqued me throughout my life. There is also Dwight Connor, the group leader who was my immediate supervisor and colleague at the National Conference of State Legislatures (NCSL). I shared with him many personal and professional ideas throughout the thirteen years I served as the ASME Fellow at NCSL. Finally, I want to thank Roop Mahajan, who encouraged me at a crucial stage of my life not to quit and made it possible for me to return to academia during the waning years of my professional life after I retired from NCSL.

On the purely professional side, there are too many people to cite by name. In a generic sense, I owe an enormous debt to all of the co-authors of the many articles we have jointly written and the books that we have co-published and co-edited over the years. Most important among this group of people is Jan F. Kreider, my former PhD student and co-author of many articles and books, Mark S. Bohn, my co-author for several editions of Principles of Heat Transfer, Yogi Goswami, my co-editor of many important handbooks, and the outstanding staff at the Solar Energy Research Institute that supported my work for a decade.

During the final stages of completing this book, I received an enormous amount of help from two students at CU. One of them is Kangqian Wu, the graduate student with whom I prepared the paper on transition, and the other is Michelle Burns, who helped me with putting together the illustrations for the book and typing parts of the manuscript. I would like to express my gratitude to Randal Fird, who graciously provided a fresh set of eyes and caught a number of errors that might have been missed. I also want to thank the professional organizations who gave me permissions to use/reproduce some articles and illustrations, and Casey Cass, photographer for the University of Colorado, Boulder for allowing me to use his photograph of the solar collectors shown on p. 273. In particular I thank NASA for the photo used on the cover. It was taken by astronaut Ron Garan on Saturday, Aug. 27, 2011 as the space station flew along a path between Rio de Janeiro, Brazil and Buenos Aires, Argentina.

Appendix I

ENGINEERING FOR STATE GOVERNMENTS

1992 Ralph Coats Roe Lecture
by
Dr. Frank Kreith, P.E.
ASME Legislative Fellow
for
Energy and Environment
National Conference of State Legislatures
Denver, CO 80202

Dedication

I want to express my heartfelt thanks to the Ralph Coats Roe Committee and to ASME for awarding me the 1992 Ralph Coats Roe Medal. When one reaches the biblical age of three score and ten, one tends to look back. The award of this medal shows that my life-long efforts to enhance public appreciation for the contributions engineering has made to the quality of our life have been recognized by my peers. This knowledge would be most gratifying any day of my life, but to be recognized today has a special meaning for me. This is the 54th anniversary of the "Kristall Nacht" when I witnessed, in my home town of Vienna, wanton destruction and human cruelty on a scale that is hard to imagine. It was the beginning of the halocaust in Europe. As a survivor of the halocaust, I have always felt a need to justify my survival, a feeling that arises probably from a mixture of guilt towards those who were not as fortunate as I, and gratitude for being alive. In this spirit, I accept the Ralph Coats Roe Medal today in memory of those six million people who might also have made a contribution to our lives if they had not perished in the holocaust.

1.   Introduction

Every generation is one of transition, but I believe that our own time portends bigger changes in the organization of the planet than we have seen since the Industrial Revolution. The Cold War has come to an end and the future strength of a nation will depend more on its economic health and political resilience, than on the number of nuclear warheads it possesses. But, the promise of the industrial age has not been fulfilled for many people and there appears to be increasing disenchantment with what technology can do to provide a better life for us and our children. Politicians, engineers and scientists promised to break the problems of human finitude, preserve the environment, travel into space, and provide energy, food, housing, education and a better standard of living for everyone. But it did not happen. Limits have popped up everywhere and people tend to blame politics and technology. I would like to share with you some thoughts on the interactions between politicians and engineers in our society and how we might improve their mutual understanding in an effort to deal with the limits that nature imposes on our expectations. Since my recent experience as ASME Fellow for Energy and the Environment at the National Conference of State Legislatures (NCSL) has been with state governments, I will emphasize their interactions with engineering technologies in my talk.

2.   Integrating Technical Knowledge and Public Policy

To solve some of the serious problems confronting our planet, ranging from protecting our environment to providing clean and affordable energy, and from maintaining a decent standard of living to assuring freedom and security, will require good engineering counsel for the political decision makers. But for the politicians to use this counsel constructively they must integrate the knowledge gained from research in science and engineering into public policy. George E. Brown, Jr., the Chair of the House Committee on Science, Space and Technology recently noted that "All the basic science funding in the world will have no positive effect on the well being of our nation if the research is not carried out within a system that can effectively digest and apply the results." Resolving our energy, waste and environmental dilemmas will require incorporating the information developed by our scientists and engineers into sound public policy.

The difficulty politicians have analyzing technical issues objectively should not come as a surprise, if we realize that only eight out of the five hundred and thirty-five current members of Congress list their occupation as engineers, a situation that is mirrored in state legislatures, where less than 1% of the 7,500 members have a technical background. Since so few political decision makers have a technical background, the majority depend on advice and counsel from experts outside the political body for their decisions.

3.   The Emerging Role of State Legislatures

In the past, decisions about major technical innovations, such as rural electrification, nuclear power, air and water pollution and waste disposal were made by the Federal Government. Congress and its staff have access to many excellent sources of information and technical advice, including the Congressional Library, the Office of Technology Assessment and the National Laboratories. For example, prior to debating the National Energy Strategy, the

DOE asked five national laboratories to prepare white papers for Congress on the subject. But, despite the excellent information pool, Congress has recently been unable to take action on many pressing problems in a timely manner. Among the reasons for this log-jam in Washington are the difficulty of achieving concensus for action when fifty states are involved, and the overlap in responsiblity among several agencies, each of which is trying to protect its own budget. As a result, state governments had to step into the breech and develop legislation to improve energy efficiency, protect the environment and manage solid waste. Some of these actions were voluntary and driven by local needs. But others were the result of federal legislation that delegated responsibility for action as well as for funding to state governments. For example, the 1992 National Energy Policy Act, which was recently signed into law by President Bush, requires state commissions that regulate electric and gas utilities to institute "least cost planning". I have worked with state governments to develop such a planning process. It requires sophisticated knowledge about energy supply options, cost effective conservation measures and a quantitative understanding of the social cost of energy. Some states have been at the forefront of trying to develop a viable least cost planning process, but none has been completely successful so far. Now all of them will have to continue this effort under a federal mandate, apparently without guidance or adequate funding from the Federal Government. Similar mandates are in the 1991 Amendments to the Clean Air Act, as well as in the bills proposed for RCRA reauthorization.

The technical resources available to a state legislature depend on its size, budget and population. They are more plentiful in a large state such as California or New York, while in small states they may only consist of a half-time staff. But, the technical resources available to any state are minute compared to those available to Congress. Traditionally, when an issue such as waste disposal or energy conservation is debated by a state legislature, spokespeople with different industrial perspectives are invited to testify and the state legislators then try to integrate the information provided to them into public policy. This approach, however, has been found wanting, as states grapple with more sophisticated technical issues than in the past.

As ASME Legislative Fellow I have been to numerous hearings before state government committees. Their first order of priority is to determine the financial and occupational ties of a witness, because legislators expect that these will color the testimony. For example, at a session on alternative transportation fuels, the first witness was an executive from General Motors who promoted the expanded use of NPVs (non-polluting vehicles), a euphemism for the electric automobile, in a large metropolitan area. When asked about the pollution from the coal fired power plants necessary to charge the cars' batteries, his response was that these plants are far away from metropolitan areas and would, therefore, not create a problem in meeting the clean air standards. The next witness proposed the expanded use of corn for the production of ethanol. When asked about the trade-off in energy between the oil required to produce the fertilizer to grow the corn and the energy content of the fuel, he responded that the oil comes from overseas while the corn is produced by local farmers who need support. The final witness was a spokesman for the Natural Gas Industry. He proposed the expanded use of compressed natural gas for transportation. It was not surprising that the legislators at the hearing were left perplexed about which of the proposed solutions was really going to benefit their state, and decided to search for assistance and information from objective sources.

From my experience at NCSL, I believe that the technical assistance legislators need can only be provided by experts who have no financial stake in the outcome of the issue and understand the political decision process. Moreover, most of the issues are not amenable to scientific analysis, but require judgement in the application of science. This, of course, is the job of engineering.

4. The Role of Engineers in Public Policy Decisions

The engineer's task in developing public policy is to provide unbiased and current technical information on various options, including their costs, risks and environmental impacts, both to the public and to political decision makers. In all political decisions there are winners and losers. Hence, trade-offs and value judgements are inexorably involved in the decision making process. The public should, of course, be heard in a democratic society, but the ultimate value judgement in a policy decision, whether to build a waste incinerator or site a landfill, whether to build a coal-fired or a nuclear power plant, whether to use available funds to improve air quality or to build better schools, rests with the political decision makers. In most cases, better information will lead to better decisions. Professional engineers can help provide technical information and recommendations in areas of their expertise to the decision makers. But to be effective, the engineer must beware of overstepping his role by trying to influence the social policy decision. The situation is similar to that of a consultant to industry: he is on tap, but not on top.

In policy decisions informed engineering judgement, not science, is involved. Policy-makers must, therefore, understand the difference between science and engineering. Science can develop deterministically and causally related understanding under controlled conditions, but the engineering application of this knowledge to a new technology under real world conditions requires judgement. Field conditions invariably produce uncertainty. No matter how well informed an engineering expert may be, he cannot make predictions with the same certainty as a scientist. The confidence in the safety and reliability of any technology increases with experience, but it can never be absolute. Consequently, there may be disagreement among experts. All interactions between technological information and political decisions involve the perception of risk and the management of uncertainty. Engineers cannot tell policy makers what to do. They can only provide options and describe relative allocations of risks.

There are, of course, occasions when scientists are asked to give testimony. In most of these situations, however, the scientist acts as a technical consultant and politicians should not expect this information to relieve them of their responsibility to make a value judgement. A story I have been told illustrating this is Senator Edmund Muskie's reaction to the testimony of a well known scientist on global warming. The scientist pointed out that on the one hand, theory predicts that the emission from more coal fired power plants would increase CO_2 in the atmosphere, trap solar radiation, and lead to a rise in global temperature. On the other hand, he said, an increase in solid particles from coal combustion could collect in clouds, reflect solar radiation, and counteract the CO_2 effect. Senator Muskie's response, reflecting disappointment in the lack of clear cut guidance from the scientist's testimony, was "Are there no one-armed scientists?"

4

5. Categories of Issues in Need of Engineering Help

Having described the role of engineering in public policy decisions, I now turn to the kind of problems in need of engineering help. In addition to responsibility transferred from Washington, the states also face technical challenges resulting from new technologies and changing social values. The kind of indigenous problems for which state governments require technical input and information fall into three general categories. The first category encompasses problems for which the scientific understanding, as well as the technology, exist for effective action by state governments. Problems in the second category occur when the physical relations between cause and effect are known, but the knowledge is inadequate for quantitative predictions necessary to take large scale action in the near term. In the third category, reliable scientific information is not lacking and the only appropriate action is to conduct research to clarify the issue.

Issues in the first category for which the technical information is adequate and new research is not likely to change the situation significantly, include energy conservation, municipal solid waste management and public transportation. In the field of energy conservation, the outcome of actions can be predicted with reasonable accuracy and the economic impact can be determined. There is still uncertainty in how people will respond to various initiatives and whether or not a legislative action should be in the form of mandates, or attempts to induce voluntary action. For example, legislation can require that the energy performance of a refrigerator be posted on each unit in a store, or it can demand that all refrigerators sold in the state must perform to a minimum energy efficiency.

In the field of municipal solid waste management we have the knowledge and capacity to institute cost effective and safe integrated waste management programs. But the fear of the public of some of the disposal methods, in particular, waste-to-energy incineration, introduces political problems for legislative action. To overcome this obstacle, it is necessary to educate legislators and involve the public from the start in the implementation of any proposed engineering solutions that may be perceived as posing a risk to the health of the public or the environment.

For problems in the second class, scientific information is sufficient for predictions based on computer models, but there are still large uncertainties in the magnitude and timing of future effects. In other words, the causal relations have been established scientifically, but the values of the coefficients relating cause and effect in the real world are still uncertain. In many of these situations, however, some state legislative actions can be recommended to ameliorate the situation, provided these measures can be justified on their own merits. Examples of these types of problems include the effect of CO_2 on global warming or the destruction of the ozone layer by CFCs. In the absence of precise knowledge on the effect of CO_2 and other gases in the atmosphere on global warming and the local climate, it is premature to build a dam to protect coastal areas from the rise in ocean levels if the ice cap should melt. But, state governments have been receptive to the recommendation that it is prudent to take measures that will reduce global warming, provided they make sense on their own account. For example, improving the energy efficiency of a state will improve its economic position, as well as reduce the production of CO_2.

The third group of problems confronting state legislatures are situations where even basic scientific understanding of the purported damage from a certain technology is lacking. These kind of situations are often driven by unfounded fears of the public or the expectation of economic gains by some industries. The best response in these cases is to develop a better knowledge base that will indicate what type of actions are appropriate. Examples of these kinds of situations are the claims that electromagnetic fields can induce cancer or that herbal medications may have adverse health impacts. In most of these problems, even though the scientific credibility of the evidence is poor and the statistical risk is low, the political impact may be enormous.

The electromagnetic field (EMF) controversy was triggered by a story in the New Yorker claiming that high tension wires could harm children living in their vicinity. The statistical evidence was largely anecdotal and extremely weak, but the outcry of a few vocal citizens lead public utilities to postpone construction of much needed transmission lines, in order to avoid the possibility of litigation at a later time. Subsequently, it was hypothesized that not only high tension wires, but also low-voltage fields from appliances such as electric blankets or refrigerators, could cause cancer. Obviously, it would be foolhardy for state legislation to forbid construction of high voltage lines now, when subsequent studies might show that it is not the high voltage line, but exposure to the fields produced by television sets that could cause cancer. Television watchers of the country, Beware!

A first step in providing assistance to legislators is to decide into which of the three categories a given issue falls, and then to provide credible evidence that can withstand criticism by the public and scrutiny by experts. This is, of course, not an easy task, but once the information is available it can be useful in other states because the emergence of concerns, fears, and desire for action on any issue seems to spread from state to state by word of mouth or by the media. Hence, the availability of a clearing house for information at NCSL is a great benefit to all state governments, particularly because they know that NCSL strives to be a credible and objective source of information and has no ties to industry or political viewpoints.

6. Some Comparisons Between State and Federal Governments

In observing the operation of state governments, I have come to the conclusion that states could become "Laboratories Of Democray" for improving the process of public policy formation. State governments can act more quickly than the Federal Government and respond to complex situations in a responsible and innovative manner, provided adequate technical information is available. It is also easier for a champion of an innovative idea in a state legislature to obtain support for enabling legislation than in Congress.

Whereas the Federal Government must deal with the heterogenous interest of fifty states, a state government serves a more homogeneous group. For example, federal legislation in the area of energy was handicapped by contradictory desires. Oil producing states such as Texas and Oklahoma wanted legislation to support drilling for more oil, while states such as Iowa and Arizona, that must import all of their oil, preferred the development of alternative fuels and solar energy. In addition to being more focused and dealing with a more homogeneous group, I think that state governments also benefit from the relative absence of PACs in their decision making process. The following examples illustrate innovations in different areas.

Risk-based Environmental Priorities

In 1990 the nation invested about $115 billion, or 1.9 percent of its GNP, in pollution control and environmental protection. State and local government accounted for about 26 percent of this expenditure. With increasingly tight budgets, many states have concluded that they must focus their environmental efforts and target the highest risks. This goal is a political challenge because "environmental risks considered most serious by the general public today are different from those considered most serious by the technical professionals charged with reducing environmental risk" according to the U.S. EPA Science Advisory Board.

The traditional regulatory approach to the environment by the Federal Government is often called "command-and-control". Pollution sources are commanded to meet discharge or emission standards by federal agencies. This is generally a rigid, piecemeal and uncoordinated approach that can inhibit the use of innovative and more cost effective pollution control. A new approach, taken by some state governments, compares the risk to human health and ecosystems from various pollution sources and targets those pollutants that are identified as the top risks. Setting risk-based priorities can help states integrate environmental and health protection programs with fiscal policies. Risk-based targeting also gives state governments an opportunity to direct their finite resources to the most pressing problems.

R and D Support by State Governments

State governments in some large and industrial states have recently begun to support research and development programs designed to help local companies and attract new industry. North Carolina's support for its industrial park is an example. In this effort the states are challenging a basic tenet of federal R and D policy, which holds that once science has discovered and explained new ideas, they will then automatically be pushed by engineering ingenuity into the market place. Economists, however, tend to rely on market pull. In the real world probably both are necessary and states seem to accept this pragmatic view.

The U.S. continues to be the world leader in basic research, but no longer leads the world in technology and manufacturing. George Brown, the Chair of the House Science, Space and Technology Committee, has proposed to provide more support for civilian technology development and to establish a Critical Technology Institute with a budget of $3-4 billion by the mid 90's. The amount does not appear an unrealistic goal for a government that spent about $68 billion for military R and D last year. But, almost all of this money was earmarked for military programs and only 4% was allocated to general science according to Jean-Claude Derian, formerly a science counsellor in the French Embassy in Washington.

Derian claims, in his recent book[1], that military R and D for the Pentagon helps only the sheltered part of technology which does not have to compete on the world market. According to critics of this policy, the U.S. needs to develop a national technology policy, with involvement from industrial users of research. This policy should set priorities and provide guidance for a transition from sheltered military research to programs in the exposed sectors that can foster the success of American industries in international competition.

In this election, the effectiveness of "trickle-down" economics has been questioned and it may also be appropriate to question the basic assumption of "trickle-down" science. In 1945, Vannevar Bush wrote a white paper claiming that funding basic research can be socially justified because its results will trickle down to applications in new technologies. This belief has led to the establishment of the National Science Foundation to fund basic research. When engineers asked for similar support to advance new technologies, the response was not the establishment of a National Engineering Foundation, but the creation of an Engineering Division within NSF, which was later elevated to a Directorate. It seems, however, that even as a Directorate, which receives about 12% of the total budget, engineering is still the stepchild in the hallways of NSF, where academic science and its peer review system set the tone. I still recall my last efforts to obtain funding for the development of a novel technology jointly with an industrial partner. The response of the NSF program manager to our proposal was, "This is great engineering, but where is the science? Without some more basic work, it will never pass peer review." My industrial partner wanted a product, not a peer reviewed paper and withdrew from the project. He recently told me that the technology is now being commercialized in another country. Several of my engineering colleagues have had similar experiences.

The criterion for state support of R and D is quite pragmatic. It simply asks, "Will the work benefit the state and create new jobs?" This may not be what academic institutions reward, but it is an approach which may be appropriate when the U.S. is trying to compete internationally and create jobs at home. For example, the state of Colorado is funding an Advanced Technology Institute designed to help commercialize the results of research conducted at the state's universities and attract new industries. Although the institutes' three million dollars annual budget is small compared to the grant money generated by some professors for basic research, the fact that the money is available without a lengthy review process and can be used to bridge the gap between research and commercial technology, makes it significant. Many other states are trying similar programs of R and D support. Although the financial resources of the states are limited, there have been some success stories which may be useful in guiding future federal research policy. To develop this policy, we should ask what kind of research is likely to improve the quality of life throughout the world, and how applied and policy research can be integrated into our national R and D agenda.

---

[1] "America's Struggle for Leadership in Technology", MIT Press.

State Energy Efficiency Programs

State legislators are keenly aware that increased energy efficiency can improve their state's economic competitiveness, enhance its environmental quality, and provide jobs for its people. The states have taken innovative approaches in energy conservation programs and some have been very successful. For example, in 1979, New York initiated an Energy Advisory Science to Industry (EASI) program to assist small and medium sized companies in identifying and implementing energy conservation opportunities. The program provides free on-site energy surveys by specially trained EASI advisors (mostly retired engineers), who work with industrial plant personnel to review utility, fuel and water-use data, and steam production needs. More than 8,300 EASI surveys were made and businesses that implemented suggested improvements typically averaged 15 percent energy-cost savings. Annual energy savings of roughly $64 million accrued in 1991 to the participating industries, and a study conducted by an independent evaluation firm showed a 20 to 1 benefit-to-cost ratio for the EASI program.

States have also done a good job of "jump starting" a variety of energy efficiency programs using both the carrot and the stick approach - incentives and mandates. Not all of them have been successful and it could be useful to ascertain the reasons for the success or failure of specific programs. This would be of great help for the National Energy Strategy to select the best programs for increasing the energy efficiency of the country.

7.   Conclusions and Recommendations

Our world is becoming increasingly complex and technological. Society and its policy makers are becoming more dependent on the operation of new and large systems whose safety and efficiency rely on engineering judgement. At the same time, the public is becoming more concerned about the risks of new technologies and tries to influence its political decision makers, often without adequate technical understanding of the issues. The interaction and linkages between the needs for energy and water, the protection of our environment and health, the management of time and waste and the desire for better education and a higher standard of living make it ever more difficult to reach consensus in public policy decisions. A majority of these decisions depend on trade-offs between initial and operating costs of complex technologies, environmental and societal impacts, pressure from different interest groups, and the public's perception of risk. Under these circumstances, governments at all levels are looking to regulation as a tool to manage technical and economic complexity as well as societal instability. All of these developments enhance the role of engineering in society. Almost every political issue has a technical component and engineers are needed as consultants and advisors to policy makers, especially in state governments. But, despite their importance as problem solvers in our technological society, engineers, unlike doctors, lawyers or scientists, curiously lack visibility. Neither does the public ask their advice, nor do the media show interest in their successes.

Obviously, technology alone cannot solve problems that have been developing over a long period of time, unless the technical solution is accepted by society. This acceptance requires a healthy interaction between engineering and public policy. For this interaction to succeed, it is necessary to:

- Recognize potential environmental and economic impacts and develop the knowledge needed to make sound predictions about these impacts.
- Educate the public on how their behavior in exercising consumer preferences and political pressures affects local and global environments.
- Take the time to interact with public action groups and inform the public of available options and their costs.
- Make a risk assessment of the options and clarify the interaction between technological solutions and social effect.
- Develop credibility for science and engineering in order that the public will trust experts to give unbiased recommendations.

New legislation dealing with technical innovations that require competent engineering advice is occurring at a time when the Federal Government is shifting more and more responsibility from Washington to the states. Hence, state governments are forced to grapple with technical issues that are new to them and require reliable information on the risks, costs, reliability and availability of technologies to produce energy, manage waste, reduce pollution, enhance competitiveness, and meet expectations of the public. Often decisions must be made to meet federal mandates or solve urgent local problems with limited budgets in a given time frame. Thus, for technical assistance to be useful, it must not only be objective and informed, but it must be available on short notice and in simple language that can be understood by people who do not have a technical education. It is in this context that the ASME Government Relations Program and the National Conference of State Legislatures provide a unique public service. Any state legislator in the country or member of his or her staff can pick up a telephone to ask for information and someone will respond within the time frame specified by the caller. I believe that this professional service has already had positive impacts on legislation dealing with waste management, energy conservation, and environmental protection. But, the scope of this service has to be expanded to meet the needs of state governments.

In addition to increasing the number and scope of technical advisors to state governments, the effective use of their counsel must be improved. To this end, engineers must work with social scientists to remove conceptual confusion regarding the separation of technical information and value judgement. These are new tasks for the engineering profession. But engineers thrive on new challenges and I feel confident that we will meet them successfully.

Appendix II

''America's hope for energy to sustain economic growth beyond the year 2000 rests in large measure on the development of renewable...energy.''

– President Jimmy Carter at SERI, May 3, 1978

Opportunities for Solar Energy in a Restructured Utility System

Dr. Frank Kreith,[1] **PE**
ASME Legislative Fellow
National Conference of State Legislatures
The Yellott Award[2] Lecture
Presented at
SOLAR 2000
Madison, WI

This is a great occasion and I want to express my deep appreciation to my peers and colleagues in the solar energy field and to the Solar Energy Division of ASME for selecting me to receive the first Yellott Award of the 21st century. In accepting this award I know that it is not given to me alone, but also to all of the outstanding researchers who worked under my guidance in building up the research facilities in solar thermal conversion, storage, thermal control of buildings, direct contact heat transfer, and solar cooling at SERI. In addition to my colleagues at SERI, also the authors and co-editors for the ASME Journal of Solar Energy Engineering have made major contributions during my 10 year tenure as the founding editor. All of these men and women achieved a worldwide reputation of high quality research for solar sciences and technologies and share in this award.

Since I retired from SERI 12 years ago, I have served as the ASME legislative Fellow for Energy and Environment at the National Conference of State Legislatures (NCSL). NCSL serves all state governments and its goals include giving timely and objective information on energy issues to all 7,464 state legislators, providing technical assistance on solar and other forms of energy to the 50 state governments, and conducting symposia on energy issues for state legislators and their staff. My talk today is based on personal reflections, observations and analyses, but does not represent an official position of any specific organization.

1 Introduction

The use of solar energy in its various forms is closely tied to the rise of civilization. But serious interest of the industrial world in solar energy began only with the Arab oil embargo in the 1970's. It is almost exactly 25 years since the U.S. Congress, in response to this "brush with crisis," passed the Solar Energy Research Development and Demonstration Act which established a solar energy research institute by Public Law 93-473. This was the first coordinated national effort for solar energy. I had been interested in solar heat for a long time and decided to accept an invitation to join the Solar Energy Research Institute as the Chief of Thermal Research when it began its operation on July 5, 1977. The high point of the Institute occurred probably a year later when President Jimmy Carter visited SERI on May 3, 1978. I still recall the President's keen interest in solar thermal conversion when I guided him through our test facilities. In connection with his visit the President proclaimed a national goal that "by the year 2000, 20% of the country's energy needs should be supplied from renewable sources." We are now in the year 2000 and unfortunately no serious observer of the solar field could claim that our efforts over the past 20 years have brought us anywhere near that goal. At present only 7% of the U.S. energy comes from renewable sources and the lion's share of that amount is conventional hydroelectric power.

There are many reasons for the failure of renewable energy to reach President Carter's goal. The price of oil plummeted after the embargo collapsed, energy efficiency measures proved effective, and nuclear power proliferated. But if we look specifically at the progress of solar programs, it has also been disappointing. I believe that the main reasons for our failure were an over-enthusiastic belief that successful solar demonstration projects could easily be transformed into marketable commercial products and a lack of realism in estimates of the total cost of heat and electricity obtained from solar systems. Both of these "faux pas" created unrealistic expectations in the mind of the American public and should serve as a warning to those who glibly expect that hydrogen will replace fossil fuels in the foreseeable future.

In addition to the above problems, long term R&D programs at SERI suffered from constant changes in direction and emphasis as DOE program managers attempted to placate the political climate in Washington. This problem was exacerbated by the fact that except for its founder, Paul Rappaport, hardly any of the directors of SERI and NREL had prior experience in solar energy development. There is no question that SERI, NREL, and other national

---

[1]Frank Kreith is the Legislative Fellow for Energy and Environment at the National Conference of State Legislatures (NCSL) where he provides assistance on solar energy, transportation and environmental protection to legislators in all fifty state governments. Prior to joining NCSL in 1988, Dr. Kreith was the Chief of Thermal Research at SERI, currently the National Renewable Energy Laboratory. During his tenure at SERI, he participated in the Presidential Domestic Energy Review, served as an advisor to the Governor of Colorado, and was editor of the ASME Journal of Solar Energy Engineering. In 1983, he received the first General Achievement Award from SERI.

From 1951 to 1977, Dr. Kreith taught at the University of California, Lehigh University and the University of Colorado. He is the author of textbooks and is the recipient of the Charles Greeley Abbot Award from ASES and the Max Jakob Award from ASME-AIChE. He received in 1992 the Ralph Coats Roe Medal from ASME for providing technical information to legislators about energy conservation and environmental protection, and in 1998 the Washington Award for "unselfish and preeminent service in advancing human progress."

[2]The John I. Yellott Award is a biannual award sponsored by the ASME Solar Energy Division. This award, in honor of the Division's first Chair, recognizes individuals who have demonstrated leadership within the Division, have a reputation for performing high-quality solar energy research and have made significant contributions to solar engineering.

al of Solar Energy Engineering   **Copyright © 2000 by ASME**   AUGUST 2000, Vol. 122

laboratories have made significant progress in basic research related to solar energy utilization. But their R&D failed in developing cost-effective measures for manufacture and maintenance of solar technologies. Furthermore, in the 1980's the administration under President Reagan drastically reduced the incentives to develop the fledgling solar technologies and market penetration of these new technologies came to a halt. But in order to avoid another energy crisis, the Federal Government initiated efforts to develop a national energy policy and in 1990 President Bush asked the national laboratories to prepare a plan that would provide a secure energy future at a reasonable cost. This effort resulted in the National Energy Policy Act, which was the last piece of legislation signed into law by President Bush on October 24, 1992.

2 The National Energy Policy Act of 1992 (NEPACT)

NEPACT contains two provisions of great importance to solar energy: The first of these deals with the electric utility energy planning process. Its goal was to place energy conservation measures on par with generation options. The process, known as "least cost energy planning," requires all states to compare the cost of meeting their energy needs from renewable sources and conservation measures with the cost of power generated from fossil fuels before building new electric power plants. This process was embraced enthusiastically by solar energy advocates because it offered all types of solar technologies an opportunity to compete on a level playing field. Furthermore, the federal government provided tax incentives to assist solar technologies to achieve market penetration and encouraged state governments to include social costs in the bidding process. Unfortunately, the implementation of least cost planning was left largely to the utilities that favored energy generation to meet the demand for an obvious reason: The profit of utilities comes from energy sold, not from energy saved.

The second part of the National Energy Policy Act of importance to solar energy requires utilities to transmit power over their transmission lines irrespective of the generation source. This requirement was strongly endorsed by utilities as well as by state legislators because they believed that competition would lead to greater efficiency, increase electric power sales as well as profits, and eventually also lower electricity costs to consumers. This process of transition from the regulated monopoly structure of the past 50 years to a more competitive system is usually called restructuring. There are opportunities in this change, but there are also risks and potential pitfalls.

In its early days, the electric industry operated without any Federal regulation. The absence of regulation led to two or three holding companies achieving dominance in the electric power market. Under this duopoly, the cost of power to consumers was high and service unreliable. Moreover, because of the expense in installing transmission lines, electric power was essentially unavailable to outlying communities. When several of the holding companies collapsed in the wake of the depression, consumers pressured the federal government to institute regulations that would insure reasonable prices and reliable service. The federal government responded to public pressure and passed in 1935 the Power Regulation Act. Under this act, public utility commissions negotiated a reasonable price of electric power for consumers and a reasonable profit for utilities. The utilities, in turn, promised to provide reliable service and build adequate backup in transmission and generation facilities to prevent undue interruptions to the consumer. This arrangement has worked quite well, but it has recently been challenged to allow increased competition in a free market for all types of power generator.

Free market advocates are not opposed to renewable energy, but they believe that solar technologies should only be deployed when they are cost effective. However, this perspective neither considers the externality and environmental costs associated with fossil fuel extraction, air pollution from burning fossil fuels, and environmental effects such as global warming and nuclear waste disposal, nor does it recognize the traditional role of the federal government to provide incentives for developing new technologies. The rationale for energy incentives over the years has been to promote a new technology during its development stages, e.g. tax incentives for oil production and R&D support for nuclear power. Furthermore, incentives have been given to pay for the difference in the value of an activity to the private sector and the value to the public sector, as for example in the rural electrification program under President Roosevelt. Today's legislators should consider that the deployment of renewable energy sources and conservation could reduce our enormous trade deficit, increase the robustness of our energy system and reduce energy-related environmental impacts. These so-called externality benefits should provide sufficient incentive to make sure that restructuring of the utility system will not diminish support for conservation measures and renewable energy. Although only few legislators in the US congress recognize these benefits, most state governments have included in their restructuring legislation provisions that preserve support for renewable energy.

3 Policy Tools to Promote Solar Energy

On a strictly economic basis, only hydroelectric and wind power generators are economically competitive with coal-fired and natural gas combined cycle power plants. Consequently, restructuring poses a threat to other renewable energy and conservation programs. Fortunately, public pressure on legislatures in state governments has led to the integration of several policies in their restructuring plans that can enhance opportunities for solar technologies. These policy tools are discussed briefly below.

Restructuring programs in some states include PORTFOLIO STANDARDS requiring that a certain amount of renewable energy, ranging from 1 to 5% of a utility's energy generation mix, be generated by renewable energy technologies. Other states have established RENEWABLE ENERGY TRUST FUNDS to be used for solar programs desired by the public. These funds are financed through a SYSTEM BENEFIT CHARGE, which is a small extra charge for electricity used that is collected from all energy customers.

In addition to these two measures which promote the development of renewable technologies, several legislative tools have been used by state governments to make renewable energy affordable to consumers. These measures include REBATES AND BUY-DOWNS which help to defray the high initial capital cost of installing solar energy systems. Japan, for example, has spurred its 70,000 Solar Roofs Program through equipment buy-downs. Several states have instituted NET METERING PROGRAMS which require a utility to purchase any excess electricity produced by small generators, such as a PV system, at retail cost rather than wholesale cost. In practice, this is accomplished by running the Watt meter, which measures the power coming to the customer, backwards when excess power is fed into the transmission grid of the utility. Thus, the customer will only be billed for the net amount of electricity used. Some states have instituted TAX INCENTIVES which include exemption from sales taxes on expensive PV systems or exemption from property taxes for wind systems that require a lot of land.

4 Green Pricing

In addition to the legislative tools to promote solar technologies, some utilities take advantage of the willingness of some consumers to pay extra for energy from renewable sources through a GREEN POWER PRICING PROGRAM. Green pricing has been given a lot of PR by utilities, but its result has been quite modest. A large percentage of the residential customers that have switched providers have opted to pay a premium to buy electricity produced by renewable technologies. But the number of residential customers that have switched is so small that the number of those that have opted for green pricing is insignificant. Moreover, only very few large industrial customers have opted for green pricing. This result is not surprising in light of a survey conducted by the National Renewable Energy Laboratory. For the average residential customer who uses about 800 kW-h of electric power per month, the monthly utility bill is between $40 and $80. The survey showed that 70% of residential customers are willing to pay $5 per month extra for green power, but only 10% are willing to pay $20 per month extra. Hence, the pool of customers that are actually willing to pay a sufficient amount for the construction of new solar generation capacity is limited. The only renewable generation technologies that can compete for ''green power'' are hydro-electric and wind in favorable locations. Since the extra payment necessary to buy PV power is greater than the entire utility bill, it is not likely that utilities will install many PV systems, except in locations where no transmission lines are available. PV is of course very attractive for countries that have no adequate power transmission system and many non-governmental agencies, such as the United Nations and church organizations, have provided funding for PV in remote areas of developing countries.

5 Evaluation of Solar Policies

In my capacity as the ASME Legislative Fellow, I have examined the successes and failures of policies in the United States and in Europe designed to promote the development of renewable energy technologies. There appear to be four key elements for successful policies:

1 Portfolio standards and renewable energy trust funds are the most important policy tools. But to achieve maximum benefit they must offer all solar technologies equal opportunity and include provisions for energy demand reduction measures in fair competition with electricity generation to meet the demand.

2 Financial measures, such as tax incentives or subsidies for renewable technologies must remain stable for several years in order to promote sufficient confidence of investors and bankers to provide loans for the implementation of these technologies.

3 Any financial subsidy must be linked to the actual production of heat or electricity rather than to the mere construction of a facility that may provide energy in the future.

4 Financial incentives alone are not sufficient for the development of renewable energy technologies. Successful policies must include a realistic assessment of the cost of the energy delivered to the consumer, assistance in maintenance of the technology once it has been built, and a continued technical evaluation of the performance of the technology in order to correct mistakes and incorporate improvements.

6 Shortcomings of Current Restructuring Legislation

From the perspective of renewable energy, there are obvious shortcomings in many current state restructuring legislations. First of all, the restructuring legislation that has been passed by most states deals only with competition in the generation of electric power, but does not consider the potential of reducing electric power consumption on par with generation as required by the 1992 National Energy Policy Act. For example, domestic and industrial solar hot water heaters and CPC driven solar cooling systems can reduce the need for electric power or heat from fossil fuels and are currently among the most cost-effective solar technologies available. Unfortunately, these options are not available under any restructuring legislation.

Secondly, most restructuring legislation assumes that competition in generation will reduce the cost of electricity to all customers. Available evidence in the USA as well as in England indicates, however, that only commercial large-scale users have benefited from restructuring, whereas the electric power bill for residential consumers has not been reduced.

Thirdly, most restructuring legislation allows utilities to charge customers in some form or other for retiring stranded assets, including nuclear power plants. This is a great incentive for utilities to support restructuring because early retiring of nuclear power plants will increase their profits, while taxpayers bear the major share of this cost. The Council of State Governements cited the nationwide cost of stranded assets as $184 billion, of which retiring of nuclear plants accounted for 43 percent or $79 billion. So far no state has required that at least a portion of this money be used to replace the nuclear capacity by renewable sources, although recent polls demonstrate that people prefer renewables and conservation to fossil and nuclear fuels by a large margin.

Finally, most restructuring legislation does not fully consider the environmental impacts and the reliability of the power system. Marketeers might find it financially attractive to buy cheap power from older, coal-fired plants in a state such as Kentucky and then transmit this power over long distances to a far away consumer where electricity is expensive, e.g., California. Such market driven decisions may minimize costs and optimize profits, but they could create air pollution, overload power lines and cause power failures.

7 Future Challenges

In closing, I want to pose three challenges that may help solar technologies achieve market penetration.

1. Since Congress under the leadership of Newt Gringrich abolished the Office of Technology Assessment in 1995, reliable and objective information about the cost and environmental impacts of energy technologies has been difficult to come by for political decision makers. After the demise of OTA, legislators have turned increasingly to the national laboratories and to industry for information about energy. Information from industry is always taken with a grain of salt and is not generally considered objective. Legislators are less aware, however, that also the testimony from national laboratories may not be unbiased. Although national laboratories have competent scientific staff, their primary goal is to secure future funding for their research projects and their employees. Scientists are neither interested nor trained to make unbiased technology assessments. I therefore want to challenge the Department of Energy to set up an energy assessment program that can provide objective and timely information on all types of energy systems, including the status and cost of renewable energy technologies. To be objective, the organization that develops this information must have no stake whatsoever in the outcome. Without unbiased and reliable information about the comparative costs of different energy choices, it is impossible to make intelligent market decisions.

About 8 years ago, the American Solar Energy Society commissioned an economic assessment of solar energy technologies. The report, *Economics of Solar Energy Technologies* , is now out of date, but it could be used as a starting point for the kind of information legislators need to chart a sensible energy future. Solar energy resources, such as sunshine and wind, fluctuate with weather and time and vary widely across the country. Consequently, also the cost of energy delivered by them is neither constant nor uniform and must be estimated on a seasonal and regional basis. The assessment should also provide recommendations for cost reductions and consider externality costs in order to give renewables a fair break in the marketplace.

2. A state legislator asked me recently in my capacity as Legislative Fellow for an estimate of the cost of photovoltaic electric power in Colorado in cents per kWh. When an NCSL intern called NREL and asked for the cost of PV power in these units to allow a comparison with the cost of traditional sources, she was told that this was not the job of a National Laboratory and was referred to companies that install PV systems. When I called one of the companies and asked for the cost of power delivered by PV in the homes that the company had equipped with PV systems, I was told that this was not a primary concern because if a potential customer had to ask how much it costs, he clearly could not afford PV. I did learn, however, that the PV cell cost was less than 50% of the total installed cost of a typical PV system. With available information on installation costs, I estimated from data in my solar engineering textbook that the delivered cost of PV electricity in Colorado is somewhere between $.25 and $.30 per kWh. Hence it seems that even if solar cells were free, PV would not be competitive in this state despite its excellent solar climate, unless the cost of deployment, installation and maintenance were drastically reduced.

There seems to be far less resistance in Congress to supporting basic research rather than R&D directed towards reducing the costs of manufacture, installation and maintenance, even though these are the main barriers to the development of competitive renewable energy markets in many instances. A study by the ARGONNE National Laboratory showed that not only has Congress cut total expenditures for renewable energy research, but those programs that have been funded mostly support university type research or basic studies at national laboratories. For example, of the $263 million fiscal 1997 budget requests of the U.S. DOE Office of Energy Efficiency and Renewable Energy, only 2% of that amount fell under the category of solar and renewable deployment and 98% for basic research. Also, the bulk of past renewable energy programs has gone to research institutions and virtually ignored the need to cut the cost of installation, deployment, and maintenance. Therefore, I challenge Congress to increase substantially the R&D budget for the deployment of systems in order to improve the competitive position of PV and other solar technologies in a free market.

3. I have previously alluded to the fact that the National Energy Policy Act contains two facets: least cost energy planning and open access to transmission lines. In order to provide opportunity for all solar technologies I want to challenge state legislators to integrate least cost energy planning into their state's overall restructuring policy. I would also urge state governments to mandate that a portion of the money given to utilities for early retirement of nuclear plants be used to build renewable solar capacity to offset the future reduction in available nuclear power.

Solar energy programs have seen many successes as well as failures in the past 25 years. We like to emphasize our successes, but we must not ignore our failures; as the great American philosopher, Yogi Berra, reminds us: "Learn from the mistakes of others; you won't live long enough to make them all yourself." With a realistic assessment of the barriers that still need to be overcome, I expect solar technologies to continue to decrease in cost and to increase in efficiency. There is no doubt that as the supply of fossil fuels dwindles, more and more renewable sources will have to be used to meet the increasing demand for energy world wide. I see solar energy as a means of helping people all over the world to obtain the energy necessary to achieve a better life and thereby contribute to world peace.

ABOUT THE AUTHOR

An escapee from Hitler's takeover of Austria, Frank Kreith is a noted author, professor, consultant, and sustainable energy expert. Frank is a valued father and friend. In his 92nd year, he is still publishing, mentoring graduate students at the University of Colorado, and providing expert reviews for proposed grants for the Department of Energy. His current project is to write, with r. Elena Virgil, a solar book for children in Spanish and English to prepare the next generation for the transition to sustainable energy.

www.ingramcontent.com/pod-product-compliance
Lightning Source LLC
Chambersburg PA
CBHW080236180526
45167CB00006B/2299